INVENTOR'S HANDBOOK

Terrence W. Fenner
and
James L. Everett

Associated Ideas Inc.

CHEMICAL PUBLISHING CO., INC · NEW YORK · 1969

Printed in the United States of America
ISBN 0-820-60381-3

Foreword

Where can I take the idea for my invention? Can I trust claims found in classified ads? Where do I go after building a workable model? How do I go about patenting, test marketing, manufacturing, and managing a business?

No book can give all the answers to specific questions; however, the authors have tried (based on actual field experience) to anticipate much of what should be known to follow intelligently the field of inventioning.

Reams of material (pamphlets, books, magazine articles) have been read and digested and over 300 years of combined experience have been compressed into these pages. The authors have tried to follow a logical step-by-step progression, leading the person with an idea or a desire to be an inventor along the rocky and many times disappointing path to successful inventing.

Two years were spent in writing this book, and over 2,000 items of mail were sent to establish lines of communication with scientists willing to review inventions, companies to work with inventors, and the other valuable listings found in the appendix. You are advised to read the entire book and later refer to specific chapters depending on the problem at hand.

This book was also written to inform people of *Associated Ideas* and its group approach to inventing and to help school future *Associated Ideas* members in all the basics of the science of inventing. Through this book an invitation is extended to join the world-wide expansion of *A.I.* called *Associated Ideas International*.

The authors wish to thank Mr. Robert Moot, U. S. Small Business Administration; Mr. Isaac Flieschmann, U. S. Patent Office; Mr. John

Maggiore and Mr. Hugh B. Sweeney, Junior Achievement, Inc.; Mr. Calvin Laiche, Vonderstein, Trombatore and Laiche; and Mr. C. Emmett Pugh, Drury, Lake, and Pugh, for their assistance in the preparation of this book, and Miss Selma Rocke, Rocke and Rhoads, for typing the manuscript and the many, many letters associated with its writing.

July 1968   *Terrence W. Fenner*
                                                          James L. Everett

Contents

your own search by mail. The professional searcher. Infringement. The patent attorney or agent. Fees. How to go about writing a patent application. Letter and petition to the Commissioner of Patents. Step by step techniques for writing a patent application.

I. The Value Of An Idea

Initiative. Direction. Energy. Assurance.

What everyman seeks, in one way or another, because he knows these will lead to a better life for himself and others. A basic concern of all—how to get them, where to find them, and most of all, how to keep them.

All of the elements that are needed are contained in one word—**IDEA!** It is the key that spells out personal happiness, because to reach a goal you will need just what it stands for—initiative, direction, energy, assurance.

Ideas have a force and life of their own. In the hands of the right person they can, and have, changed the face of the world. Just one man can build or destroy a whole civilization—be he a Caesar, a Napoleon, a Washington, an Einstein.

Every person who lives changes the world around him and affects it in some way. It doesn't take much, even a baby can do it. Countless stories of happiness and tragedy have been written just to show the influence of one child.

It is a kind of immortality that man seeks to leave his mark on the world, whether through children, the way he controls others, or the work he accomplishes that may leave a lasting monument to his name.

Long before the sophisticated probings of the mind began through psychiatry, the ancients had uncovered the secrets of the power of the mind. They knew that an idea, of its very nature, draws from the subconscious sufficient force to carry it through to completion. It is a teleological principle of the mind, seen reflected again and again in natural processes as minute seeds grow to ordered maturity, each according to its own pattern, acorns into oaks, sperm and ovum into men.

Our newest science, parasychology, is authenticating the strange powers of the mind, and is developing a whole new language to explain mind-reading, power over matter, telekinesis.

1

We are just recently beginning to find explanations and verification for the fantastic feats of the oriental yogis and fakirs who can be buried alive for days without harm, seemingly know things happening in other parts of the world with no known method of communication possible, and live to fantastic ages with mental alertness and health.

Yet Yoga, in its true sense, is a religion, a way of life almost impossible for the Western mind to follow because we are so oriented to living in the hustle of civilization. But as a coherent system of knowledge more than 5,000 years old, it has something to offer that bears investigating.

Even the first statement of Pantajali in the *Sutras*, which make up the bible of Yoga, makes sense for modern man: "To become aware of yourself, there must be a complete mastery of the mind and emotions." One can then develop distinctive, individual conclusions based upon a true picture of the facts, instead of a jumble of confused impressions, half-thoughts, and values that are merely a projection of the surrounding world.

Freud attacked the problem in a different manner, but shed a critical, scientific light on the murky world of the subconscious. He demonstrated that most of our drives and cravings are merely manifestations of basic instincts for self-preservation and self-expression—the ego and the id battling for control.

These are powerful forces which, if unleashed, can almost take over control of a person and drive him to action. The energy is there, and we spend a great portion of our lives learning how to control it.

The almost frightening power of the mind is well demonstrated under hypnosis, when direct access to the subconscious is gained. In a deep hypnotic trance a person can remember every detail in his early childhood, though it might have happened 50 years earlier; a small woman can be made stiff as a board, capable of supporting great weight placed upon her body; subjects can be made immune to pain, even bear a child at a precise hour and minute without pain—a feat that requires control of every part of a woman's body.

But if so much ability is there, why doesn't it come to the surface naturally, to be put to use more easily?

Because it has no direction, it is force without a channel, it waits for a well-defined command to put order into the chaos of conflicting drives, feelings, instincts, half-thoughts. It waits for precisely—an *idea.*

An idea of sufficient importance can motivate a person to change his whole way of life. Forces will be gathered and used to seek out the needed solution. And a calm, deliberateness of purpose will replace confusion and frustration. Assurance will grow as the confidence of power becomes stronger and stronger. The world looks for a man with an idea.

In a very real sense, it is the basis of our economy, this business of ideas. The United States has been called one of the most creative nations on the face of the earth, and because of our inventiveness (with, of course, the resources to work with) we have become the richest country in the world.

Our way of life depends upon new products every year. It is the basis for growth, for keeping the gross national product at a high level to fight recession. An ever increasing population with more and more money to spend demands to be cared for efficiently, so that living standards may rise. A growing labor force creates social tensions that become political issues as the government tries to find new jobs for those displaced by technological improvement. Capital must be invested and give reasonable profit to keep the economy moving forward. Even Russia watches our rate of growth closely, hoping to match it and declare its system superior to ours.

The basis of sound economic growth can not be found in the statistics of yearly steel production or the number and variety of new cars, television sets, etc. It depends upon new developments in science and technology, innovations and new processes that will form the basis for capital investment. There must be an increase in the effective use of the underlying ideas and skills of the nation.

Business and government realize that research is fundamental to our growth pattern. Both are seeking ways now to systematize innovation —to create whole new industries that cooperate in the production of basic inventions.

Vice-President Hubert H. Humphrey in the June 1966 issue of *Popular Science Magazine* tells how important inventors are in keeping U.S. business healthy and dynamic. He points out that:

> Independent inventors accounted for 40 of 61 important inventions made since 1900. This was the conclusion of a study cited in a classic analysis, *The Sources of Invention*. Without the support of organized laboratories, these independent inventors changed scientific and technical history.
> Times have changed. Development of an invention often requires massive team effort and sizable sustained investment. But the basic idea of an invention—the original concept—and at least its early stages of realization, can still be the province of brilliant lone inventors.

Where developments once grew almost by chance—like the automobile, railroads, electricity, telephones—now it is recognized that if we are to keep the forces of our economy balanced we must actively seek out and support new approaches in science. From this outlook have grown whole new fields of business endeavors: the manufacture of transistors, aerospace flight, computerized data processing, the laser beam. It has also made research (both government and private) one of the largest industries in the United States, amounting to more than $15 billion dollars a year.

But though there has been a 3000 percent rise in industrial research in the past 35 years, there has been an increase of less than 5 percent in patentable inventions. This, though evidence indicates that the income realized from an average U. S. patented invention may exceed $1 million, and U.S. corporate revenues from licensing are estimated at more than $300 million annually.

What is desperately needed is a rise in new, patentable inventions.

Without our trying to give a definitive answer as to why a greater number of scientists and research teams are not producing more patents, one point seems apparent. There must be more attention given to the creative resources of the largest part of our working force—the non-scientist and engineer. The majority of these people are technically trained in some aspect, and have an advantage over the theoretician in that their ideas are usually more practical and capable of commercial development.

The people who deal in the manufacturing, distribution, and sales of goods and services know intimately the inadequacies and vexations of the items they work with. John Doe working on the production line can often come up with a simple answer to a problem that stymies the experts because they are too far away from the day-to-day mechanical details involved. (The fact that John Doe sometimes sits on information is a reflection of poor organizational communication. Often he is simply not asked, or not given enough encouragement and assurance to speak up, or made to feel needed so that he too tries to work out his part in the overall problem.)

Many companies recognize the gold mine of latent talent working for them and have instituted formal programs of recognition to reward employees that come up with a new plan or idea. It always meant favor to an employee to give an "extra hand" to his company, even via the suggestion box. Sometimes it would be the deciding factor in a raise or promotion. But all too often, there seemed to be no immediate payment or recompense for a profitable idea.

Now, literally hundreds of companies give cash awards ranging from $25 up to tens of thousands of dollars. International Business Machines (IBM) has paid as much as $40,000 to a single inventor. Westinghouse gives a healthy percentage of what is saved or made on a product to the inventor.

The government has a standard program to encourage employees' suggestions for ways to save money and operate more efficiently. With approximately 1 out of 10 persons working in federal jobs there are millions who can put money in their pockets without going through a long, involved patent application—just by paying extra attention to what they are doing.

Taking just one year, 1961 for example, here is what was included in the Government Employees Incentive Awards Program:

Value of measurable benefits	Number of adopted suggestions	Amount of awards
$63,927,159	$110,295	$2,669,998

$485,000 was gained in material savings in the manufacture of the Polaris missile through interchangeability of missile parts by the work of Walter P. Moore, an engineer with the Bureau of Naval Weapons in Pomona, Calif.

$302,800 in direct savings was realized by the Agriculture Community Credit Corp. through special efforts of seven employees of the Dallas, Texas office by developing an improved method of fumigating stored milled rice under polyethylene tarps.

$166,000 reduction in man hour costs was saved by using an inexpensive training aid for instructing students in maintenance and repair of radio equipment. The idea was proposed by Robert J. Hornbeck at Fort Sill, Okla.

Here is the award scale for tangible savings under the Incentive Awards Program:

Savings	Amount of the Award
$1 - $200	$10
$201 - $1,000	$10 for the first $200 in savings and $5 for each additional $100 or fraction thereof
$1,001 - $10,000	$50 for the first $1,000 in savings and $5 for each additional $200 or fraction thereof

| $10,001 - $100,000 | $275 for the first $10,000 in savings and $5 for each additional $1,000 or fraction thereof |
| $100,001 or more | $725 for the first $100,000 in savings and $5 for each additional $5,000 or fraction thereof. The maximum award for any one contribution is $25,000 |

The opportunity for invention is open to everyone—women definitely included. It takes no formal schooling or post-graduate degree in mechanical engineering to know the many practical household problems a woman deals with. There are popular, widely syndicated columns in most large daily newspapers that are devoted solely to answering these problems or trading information on how to make housework easier, revealing needs common to thousands of women. Often the readers write back to the columnist offering ingenious solutions using makeshift apparatus already found in the home. Various science magazines pay (albeit a small amount) for suggestions about what kind of gadget homeowners would like to see on the market. The editors receive an overabundance of such suggestions and, based on their years of experience, choose the ones for publication that they believe reflect the needs of the majority of the people. They are showing a ready-made market, just waiting to be tapped.

With an inquiring eye and a dogged persistence, the reader should be able to come up with a score of practical, patentable inventions. Less than five years ago, *This Week* magazine asked readers for fresh ideas for inventions and got a barrelful. They were things people had known about for years which, if they had followed up themselves, might have meant substantial material rewards.

Here are just a few of the suggestions already on the market or waiting to come out:

Artificial, but realistic lawns

Triangular mops and brushes for cleaning corners

Plastic throw-away window shades

Shut-off for telephone ring when not wanting to be disturbed

Television earphones so that one person can look and listen without disturbing others

Disposable pocket-size raincoats and rain hats

Hot aerosol shaving lather

Lifelike, durable artificial shrubbery

Retractable seat belts

One of these might have been your idea five years ago. If you didn't use it, might it not be time to ask yourself "Why did I let it go to waste?"

In the more technical business world the approach will have to be different. It is rare that an outsider can provide a valuable contribution to a field that he knows nothing about. People who have been working in a specific endeavor for years most often have tried and discarded, for one reason or another, simple technical changes in what they are doing.

The "obviousness" of an invention (a mere change in size or shape, an aggregation of previously know devices, isomers of existing compounds) is one of the first factors to be considered. In other words, would an expert, working in the field for years, have thought of this process or product? If the answer is an obvious yes, then you may feel fairly certain that there is either an existing patent on it, or it is not novel enough to warrant a patent.

However, if you find unusual results following known procedures, or much more efficient results than were realized before, you could very well have something worth following up and should go a step further and have it evaluated by a professional in the business. (How to protect your idea when showing it to another person is covered Chapter V.)

Of special concern to those working in the chemical field is a fairly recent decision by the Patent Office to allow a claim for solutions reached by repeated trial and error over a long period of time. This discarding of unsuccessful methods or approaches that finally ends in tangible results is definitely patentable because the process shows an advance of knowledge and usefulness as it enables one not to have to repeat the same time-consuming and costly mistakes to arrive at a solution.

Mistakes do not always mean failure. There is always something worth salvaging in the original project, even though it may prove not worthwhile.

Some of these cases of failure that turned into fortune almost have a serendipity aspect to them. One of particular interest concerns the Kalvar corporation, a then small company operating out of New Orleans. They were looking for an improved printing plate and had hired a research firm to investigate the possibilities along the line they were seeking. But the firm came up with negative results.

However, the laboratory held a conference with Kalvar to prove that they had actually performed a thorough investigation. Kalvar brought along Dr. Robert Neiset, head of the physics department at

Tulane University, to verify the findings. As stated, the proposed printing plate was not workable, but he discovered in the process a whole new concept of film developing that used heat instead of chemicals. Under his guidance, Kalvar branched out into an entirely different field so lucrative that the company's stock at one time went from $1 a share up to $450 a share. It has since leveled off to about $200 a share, but the company is well founded with contracts with the military and certain selected industries.

With all the new leisure paraphernalia on the market, the ever-growing fields in plastics, electronics, synthetic fibers, and the multitude of simple mechanical gadgets that people are willing to try (sometimes just for the fun of it), you probably already have thought of one or several improvements for existing products that could be profitable.

But before you go any further, there is a very practical rule to be followed. It is something that should be done at the earliest possible time, even when your idea is still in the formative stages. In fact, it constitutes you're only recognizable record of claim should anyone try to exploit your invention or say they had it first. (In patent suit cases it is called infringement and happens to only about one out of 50 patents.)

At the very beginning, and continuing all the way through to the completion of your invention, keep a written record of just what you have done. There are several "must" conditions that have to be complied with in order that it may be a legal document.

1. The notebook must be bound in some manner (cloth-backed, glued, etc.), not just a loose-leaf filler notebook. This is done to prove that new pages were not inserted at a later time and back-dated to incorporate information that wasn't known at the time.

2. To further protect yourself, the pages are to be consecutively numbered and *dated* and *signed*, with sufficient space left at the bottom of the page to have it witnessed by two people. This does not mean notarized, only shown to two reliable people, not related to you, who authenticate the fact that the entry was made by you on such and such a date. They do not have to witness the page on the same day as the entry was made, but it should be done within a reasonable length of time—say two weeks, not more than a month.

3. The witnesses mentioned above cannot be a wife or husband (neither can testify for or against each other) nor can they be close relatives—for instance, a brother-in-law, first cousin, son, daughter, etc. Testimony of close relatives has already been disallowed in patent suits, so the precedent has been set.

4. As your work progresses to more and more refined detail so that the idea is reduced to practical application or a model, different people can witness the pages at different times, though ideally it should be the same person. The reason behind this is not only to have someone reliable that you can trust, but also that he be able to show that he understands your invention and how and why it works. He does not have to be an expert in the field, but the mechanics have to be clear to him. For your protection, the Patent Office has declared that the witness can not be a co-inventor or sole inventor should he try to steal the idea.

Demonstrating the workability of an invention for a witness can be made simple. For instance, the mechanics of a machine developed by Joe Budde to expose 1500 film frames a minute for printing was proved sound by attaching negatives to a spinning turntable and then developing them in sequence by means of a strob light. The commercially developed machine might be very complicated, but its underlying principle was easy to demonstrate.

5. It would be best to make the writing in ink for permanence. Also, do not erase errors, wrong conclusions, blind alley approaches, etc. This shows that you continued working on the idea with reasonable diligence, (it could take two or three years to perfect some inventions but there should not be too long a lapse of time in entries) and also that you have not tried to deface the copy to put in information not known at the time. Instead of erasing, cross through the minor corrections if you wish, but then initial the corrections.

6. Be specific. If you are working on a chemical formula, list all the chemicals used, how they were used, and the results obtained. Don't hold back any "secret" parts from your witnesses for fear they might steal the idea. If you follow these steps you have absolutely nothing to fear from someone's stealing your invention because you have all the proof that is necessary to show priority of your idea contained within the notebook. But if you neglect to demonstrate a crucial step, then your witness cannot say he really understood the nature of your invention.

When you buy your notebook for recording the description of your inventions, you might consider pasting the sales slip onto the inside front cover of the notebook. The sales slip should show the date the book was purchased. More important, you could have your first description notarized. Likewise, it would be adviseable to notarize a description every six months. The notary could state that every page of your notebook up to that date was complete or lined out (both front and back).

The first person that attempts to patent an invention is not necessarily the one who gets it. This is determined (in case it ever did come to court) by proof of who had the *idea* first—and the notebook will be proof of that. Of course, work on the invention has to have been continued so that a claim of abandonment can not be made.

There have been many famous inventions that were thought of and worked out at the same time, but for which the patents were awarded to the first person who made a detailed record of it. The automobile and the radio are cases in point.

A question many first-time inventors invariably ask at this stage is "Who can I trust to explain my ideas to?" The realization that they have to trust *someone*, if only to protect themselves, comes hard to many people. Patent attorneys are long familiar with the inordinate secretiveness of some of their clients, often to the detriment of the inventor's progress. It's a big "hump" for some to get over, and yet more than any other single thing it can spell out the difference between years of fruitless effort and continued success.

If there are any so-called secrets that separate the professional from the amateur, one of them is the knowledge that you must work with and get help from someone else. No one can teach you creativeness, or how to find a need, or give you ideas, but in the very complicated business of patenting and marketing an invention you will have to work with many other people to succeed.

Finding those persons who will be of benefit to you is well worth your time and effort. Caution should be exercised, however, not only to seek out those you can trust, but more important your associates must be able to appreciate what you are trying to accomplish—who have knowledge in the field, are businesslike in their approach and level-headed in their suggestions.

Close friends are not always the best judge of the worth of an idea. It is often very difficult for them to be unbiased. They quite naturally would like to see you succeed, and may be afraid to point out obvious flaws. Friends will give the necessary moral support but not the professional knowledge you need.

You probably already have discussed your plans with someone, but here are some suggestions for additional help that may be of benefit (that is, before the last stage when a patent is about to be applied for).

1. On p.295 is a list of scientists, university professors, and professional research and testing laboratories that will review and evaluate

your idea for $10. They are not in a position to judge patentability (only the Patent Office itself can determine this) but they will be able to point out flaws, show what improvements can be made, or even suggest different approaches which might prove more practical or productive. These people have all been contacted personally and have indicated their desire to help cooperate with new inventors. They won't test market your invention or do sample surveys, extensive comparative chemical analysis, etc., but they will give you an analysis that could cost up to $50 from some firms that offer the same service.

2. You will also find on p.297 a list of inventors (along with their fields and major interests) who are willing to help others get started or collaborate and pool resources and knowledge for specific inventions. They are scattered all over the country and by writing them you should be able to find someone close enough to your locality to work hand-in-hand with for each other's advancement.

3. Within your own town or city you can find intelligent, success-ful people who will be glad to "give you a push." Inventors, like other creative people (writers, artists designers, etc.), often have to work alone on their project and are only too happy to discuss mutual problems with someone else who is sincerely interested in their same occupa-tion. As there is no formal organization for creative workers they must depend upon personal contacts with others like themselves to continue "apprenticeship" in a business where the learning process never stops.

Where to start looking? You could run an inexpensive classified advertisement in the business personal column of your local paper. The ad could be worded like this:

> Inventor would like help from someone in chemical field to reduce
> idea to finished product. Name, address, phone number.

More specifically, here are 25 sources you could try:

Bankers
Lawyers
College professors
Chamber of Commerce
Industrial development organizations in the community
The Commerce and Industry agency in your state
Business and Defense Services agency U.S. Department of Commerce, Wash-ington, D. C.
U. S. Small Business Administration
Consulting engineers, chemists, etc.
Better Business Bureau as to the availa-bility and reputation of local companies
National Inventors Council of the U. S. Department of Commerce, Washington, D.C.
Local office of the U. S. Department of Commerce

For military iventions consult the arm-ed services	Your employer
Other inventors	Your friends
Small Business Investment companies	Your local governments
Experienced businessmen in the com-munity	Trade associations
	Labor unions
Real estate agents	Rotary Club
Accountants	Companies in your area

4.   A good source to turn to when having a model made or a chemical experiment run is to contact a company that manufactures the material for your model or chemical invention.   Often these companies will either run the experiment for you (at no charge) or suggest where you could go to have it accomplished.   Patent law holds that these persons and/or companies are merely "a pair of hands" for the inventor and hence you would not lose patentability even though you did not perform the actual reduction to practice yourself.

5.   If you have the time and interest, promote an "idea" club of your own.   This could be composed of only a few people on an informal basis, or you could go so far as to set up an incorporated research organization such as "Associated Ideas," which not only helps the members, but hires out as a consultative service to individuals and businesses.

The benefits you will reap from these individuals and businesses or groups are derived from years and years of experience, hours spent experimenting (sometimes up blind alleys), mistakes learned the hard way, money spent for extensive technical libraries.   In "Associated Ideas" alone, there are over 250 years of combined experience in a dozen professions.

In such a wide-open, unstructured field as inventing, it would be less than reasonable not to take advantage of such an opportunity to realize the short cuts to success.

Before one advances from the field of ideas to that of inventions, mention should be made of the worth of non-patentable projects such as merchandising ideas or improved business practices submitted to a company for consideration.   This might be a cheap advertising gimmick or stunt, a better method of distribution, discovery of a new use for an old patent that is now in the public domain, etc.   You are, in effect, selling knowledge like a doctor or lawyer; knowledge that is "available," but is of use only when applied correctly in a specific circumstance.

The courts have ruled that in order for a person to collect payment from a company for an idea it is necessary that a common law contractual agreement be executed between the company and that person.   This holds true for both patentable and non-patentable cases, i.e. anything that is still just in the idea stage.

For your protection, a very definite sequence must be followed (many major companies won't even consider an idea without proper safeguards for both themselves and the submitter). First, write the company explaining that you think you have an idea for an improved business practice or the like, but do *not* disclose the idea. Instead, ask the company for a disclosure form on which to submit the idea, with a duplicate copy for your records. (Companies that consider and use unsolicited ideas have their own disclosure form which must be used. However, if one is not available or if you would like to compare forms with a fairly standard contract distributed by the Small Business Administration, you will find one on p.293.) When the form arrives, and it may take a few weeks, fill it in completely and state your idea in as much detail as possible. If only a single form arrives, obtain a photostat of your completed form for yourself. In addition, keep a record of all your correspondence with the business firm, for it may one day forget about your original disclosure and use your idea. A simple reminder of your previously submitted suggestion may prove very valuable at this time.

If the company still ignores your demands for payment, and if the idea is of sufficient monetary value, a lawyer should be consulted. If you have kept records of all your correspondence, a lawyer could probably be persuaded to accept the case on a contingency basis. Remember, however, that you cannot submit an idea in complete form when it is unsolicited and still keep claim to it because you have not entered into a contractual agreement. Also, you are not allowed to broadcast your idea in any way to some other person under any condition other than obligation to review in confidence. If your idea becomes public knowledge, then your claim as first and true originator is lost.

Helping to promote and market ideas is a new club which, if it fulfills its promises, is worth investigating. It is entitled "The Idea Club of America," Inc., and its address is P. O. Box 14, Newton Centre, Mass. At this writing (1968) no conclusive proof of its integrity of claims is known, but it appears to fill a definite need in a legitimate, straightforward manner.

For a $5 yearly membership and a $2 fee for each idea submitted, the club will evaluate the idea and, if it has merit, will contact a minimum of five companies that may be interested in purchasing the idea or invention. If the idea is not considered worthwhile or original, the club notifies the submitter and informs him of the evaluation. The $2 fee is retained but is applicable to the next idea submitted. Out of 21 ideas sent to the club on four different occasions, the organization declined to list five, but made fees payable for further submissions.

If ideas come to you naturally you might consider selling some of the best ones at a very reasonable price to build a reputation. Businesses that see you constantly submitting good ideas may want to hire you or seek out your advice. By building up a clientele you could gradually raise consulting fees and later open an office and enjoy doing what comes naturally.

A profession similar to inventing, in that is has a common problem of directing creative endeavor into a highly structured commercial market, is writing. The mechanics of writing can be taught, but no one can teach a person how to inject that "spark of life" in fiction, that ability to describe events meaningfully, symbolically through the vision of his own personality. But, in the very broad yet diverse field of the printed word, necessary requirements must be met to satisfy particular readership demands.

Writers are somewhat in advance of professional inventors in forming national organizations, schools, and criticism services that serve as a means of apprenticeship. They know the necessity of like individuals' banding together to pool common knowledge, and have worked out effective systems for this purpose. Many, many persons with potential for writing have had article after article rejected because they weren't slanting it to the right market, didn't know the particular form of example or general statement that some magazines want, hadn't even researched the publishing field to know whom they should submit to— and then became very successful with the same articles that were rejected before, because a writer or team of writers showed them what rules must be followed, what formats are expected in certain publications.

Such a systematized school of knowledge does not exist yet for inventors, at least not on the national scene, but the method of apprenticeship training, albeit on a smaller scale, has been worked out successfully for inventors. The approach is very similar in application to the one writers use, with the exception that you have to contact two or three different sources to get complete assessment of your work, i.e. idea evaluation, reduction to model, manufacturing.

Inventor's Handbook, in attempting to fulfill this gap, will give all necessary steps with detailed references to sources of additional help for testing, marketing, etc. so that it is possible to carry your idea through to completion on an individual basis. However, just as writers have found out through long, hard experience, it is highly recommended that you work with someone else or some group to gain the advantage of supplemental knowledge and experience.

In a capsule version here is the procedure to be followed to develop an idea. It is arranged as a flow chart to give a clear, overall picture for easy reference.

a) Write your idea down on paper. Let the thoughts flow. When you can no longer think of something new, reread and correct the material. Then rewrite the idea, possibly changing the order of your material. Sign and date your writings and put them in a hard bound notebook with consecutively numbered pages (use every page). Have this notebook witnessed by two people for each entry. Preferably select witnesses who can understand the idea. Have them sign their names and make a notation of the date.

Then go to your city library to get all the available information on your idea. There may be answers already worked out to cover weak points or problems in your invention.

b) Make a model. It does not have to work perfectly. A model will often reveal improvements and patentable features of your invention. Put the details and possibly a picture or diagram of the model in your notebook. Sign your name and the date and have the model and notebook witnessed by two people who then sign their names and the date and state that a model was made and they understand how it works. In the field of patent law it is very important to reduce your idea to practice.

c) Get an opinion of the worth of your invention. Contact someone who will be unbiased, not a relative or friend.

d) Check to see if your invention has already been patented by making or having a patent search made. (Both approaches will be covered in detail in Chapter V.)

If your invention has not been patented, file a patent application for your protection.

e) Test market your idea. This will iron out the bugs and possibly show faults that need to be overcome. If the market test is successful, this will greatly increase the value of your invention should you decide to sell the rights to a company.

f) Decide whether to try to sell the invention to someone or market it yourself (possibly part-time without giving up your job).

```
┌─────────────────────────┐
│   Idea For An Invention  │
└─────────────────────────┘
```

Make Or Have Model Made To Test Idea

1. Names and addresses of commercial research and development companies available for hire
2. Names and addresses of inventors with skill and experience for collaboration (college professors)

File A Patent Application

1. How to search U.S. Government patents and write your own patent application, *or*
2. Names and addresses of the Washington patent attorneys and agents

```
┌─────────────┐
│ Test Market │
└─────────────┘
```

Sell The Right To The Invention

1. Patent brokers
2. How to sell invention, *or* companies wanting inventions
3. Small Business Administration aids to inventors

Manufacture the Invention Yourself

1. Investment companies
2. How to manufacture and distribute yourself
3. Small Business Administration aids to business

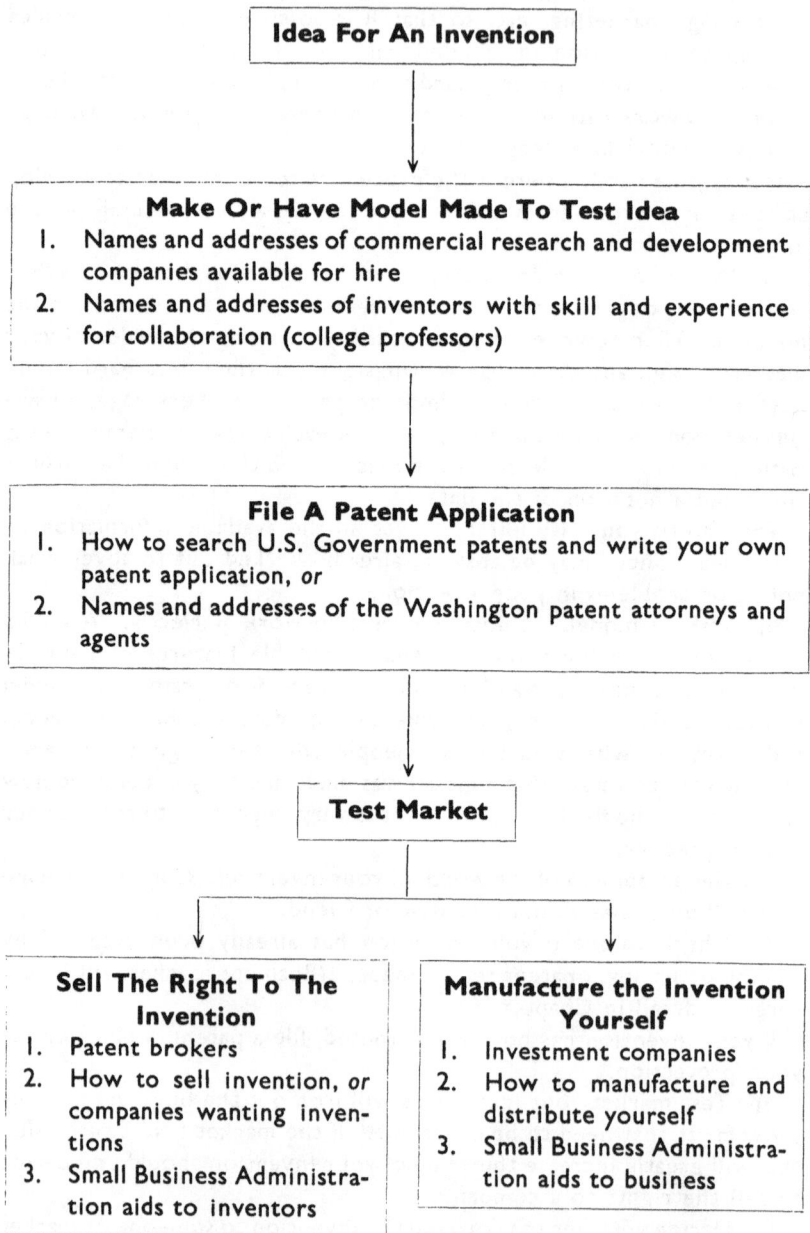

Figure 1.—Developing an idea.

II. The Most Important Step

Everyone is an inventor. The housewife, though chided for thinking she can do wonders with a safety pin, often comes up with ingenious answers to a multiplicity of practical problems. Who has not heard of the marvelous uses for a piece of baling wire, dating back to the days of the Model T Ford? It may be makeshift, but it shows an awareness to seek new uses for familiar products, to change the environment for our needs. The principle of creative inventiveness underlies this answer-seeking approach and those who take the time and trouble to perfect their ideas are rewarded by an eager consumer market.

Take, for instance, the simple invention of a negative-roll film dryer that came out recently. Photographers have long been handicapped in this one step when developing film because it takes up to 45 minutes for the film to dry before being able to print. Some make their own film dryers—consisting essentially of a hand-held portable woman's hair dryer mounted inside a suitable box to hold the film strips. In ten minutes, they have dust-free dried film ready to work with. Now, with only very few modifications for freer flow of air, smaller case dimensions, and better heat control, a machine is being manufactured for retail sale at $90 and up. It is a profitable item, and at one time had a waiting list of prospective buyers. What is amazing is that many photographers had constructed their own handmade models, knew the widespread demand for such a machine, and yet never bothered to investigate the patent possibilities.

An even more obvious example of a need by those involved with a problem is a revolutionary fish scaler based on a principle that is literally hundreds of years old. With it, one person can easily clean 45 eight to ten pound sheepheads in 20 minutes, with no flying scales and no mechanical power source. A fisherman's dream come true!

It all began somewhere back in time with the Choctaw Indians, a cross-breed race living in the lower part of Louisiana.   Their livelihood was fishing, and through handling the rough lines, paddling, and pulling in nets they developed very strong, deeply calloused hands.   They found that by starting at the tail of a fisn they could insert their fingers (with the nail side down) undneath the scales and run their hand right up the fish, peeling off scales with a minimum of effort.

The man who capitalized on the method was Dr. Arthur L. Robichaux, a dentist who has been fishing for 45 years and all that time hated to clean fish.   He had tried various methods, but most of the time relied on a hatchet, which was dangerous, inefficient, and scattered scales everywhere.

One day a patient of his, Bootsie Caballero, told Dr. Robichaux he saw the Indians off Grand Isle skin fish with their hands, an idea so outlandish the dentist had to investigate it in person.

The next time he went fishing, they showed him how to do it, and he was flabbergasted at the sheer simplicity and ease of the operation. He flipped the scales off slowly and carefully at first, just as he was told to do.   But, when he got home, he wanted to show off, and tried to go as fast as he could.   The result—he cut open every finger of his right hand on the razor-sharp scales.   He had forgotten how tough the Indians' hands were and how careful they were to place their fingers just right so that they would not slip.

Being a dentist he found himself out of work for two weeks until his his hand healed.   Instead cf mourning about his bad luck, he took the time to examine the problem and realized that a simple wedge-shaped device with a handle could serve in place of the fingers to eliminate danger.

His first model was made out of the plastic that is used as a gum base for false teeth plates.   It was pliable and strong enough when hardened to practice on.   The next step was a mold, improvement in design, and the use of a product called Plastic Steel that was capable of holding a sharp edge and was generally stronger in all aspects.   With the continual urging of friends who wanted the device for themselves, he was soon in business, though it actually took 15 models and months of testing before he was satisfied that the tool was perfected.

Another case of an independent inventor who was not content with things as they were is even more spectacular in its success story, with returns topping $25 million!   His name is Chester Carlson, a graduate of Caltech who went to work as a researcher for Bell Telephone Laboratories.   But he found he was not cut out for the "team approach"

Aug. 14, 1962   A. L. ROBICHAUX   3,048,884

FISH SCALER

Filed April 5, 1960

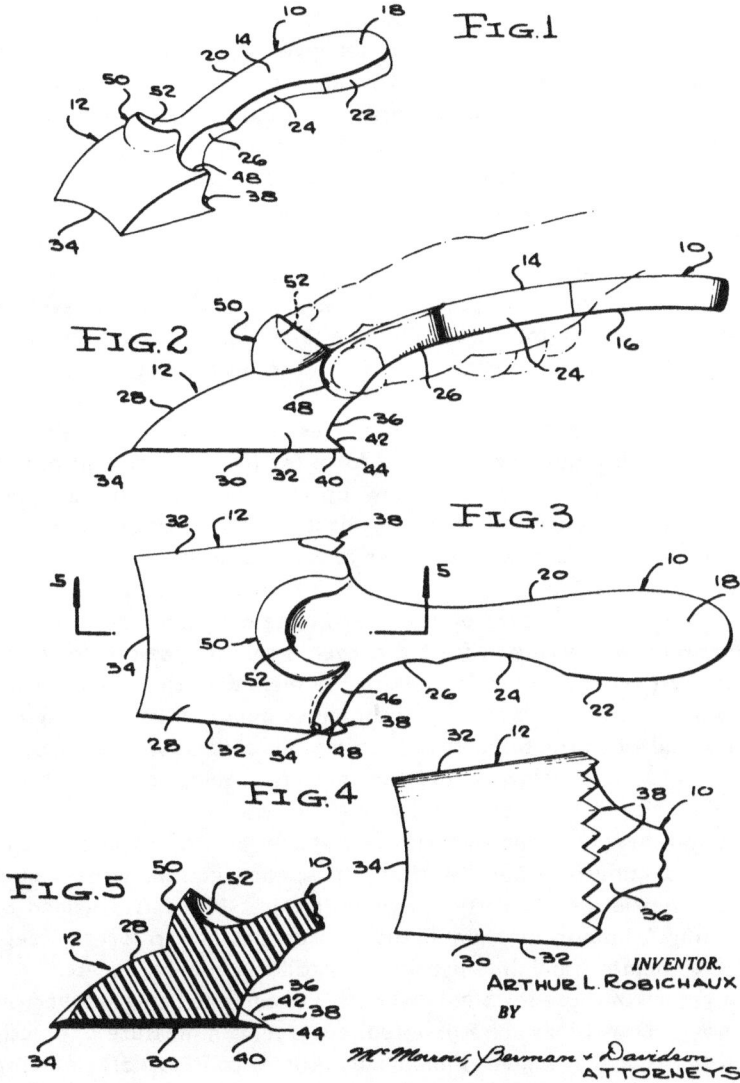

Figure 2. Robichaux fish scaler.

and gave it up to become a patent lawyer, practicing in Rochester, N.Y.

Working with patents, and all the copying of documents that is part of the process, he kept wondering whether there was a cheaper, easier way to do it. He lived in the home city of the giant Kodak laboratory and knew it had most of the photographic-chemical processes already covered by patents. But he would not give up the nagging idea that there had to be an easier way. One day it hit him: Use statically charged paper and the ink or carbon particles would jump to the electric charge pattern on the paper. So, working in his kitchen with crude equipment and infinite patience, he evolved the formula for xerox copiers, a mainstay of the business world today.

One can ask whether there is a common principle in the many examples that can be given of someone suddenly coming up with a new idea that just seemed to "pop into his head." How can the average person, faced with a job that makes the same demands day after day, see things in a new light, shake up his mind to let his imagination flow free?

There is a way to do it, it even has rules already tested and proven effective to bring out new ideas, to form intriguing juxtapositions of familiar objects that lead off to new fields. The principle has been called by many names, but the one most widely accepted now was coined by Alex F. Osborne; he speaks of it as "brainstorming."

Brainstorming—a beautifully descriptive term for what actually happens: letting loose a storm of new ideas, opening the windows to the imagination and drawing in fresh air, even a gale if necessary to clean out the musty corners of stale thinking. A mental earthquake is necessary to open up channels to the subconscious mind where the imagination lies, and when the tempest is over all kinds of ideas will flow freely; some good, some irrelevant, but the important thing is letting them flow. Criticism is for another stage in the operation.

To help renew the childish curiosity that sees everything as possible, think for a minute of the "wildest" in science fiction, what would make the world a truly better place to live in. Man has dreamed of many things—immortality, unlimited pleasure, ability to forego sleep, increased memory and intelligence, controlled child birth, etc.

And yet, all of these so-called "wildest" dreams are being worked on right now. Drawbacks are still noted, but a certain measure of practical reality has been reached. Immortality is being considered; by putting a person in a deep freeze, it is claimed that one can be perfectly preserved with all bodily processes stopped, and revived hundreds of years later when technology has advanced enough to lengthen life span immeasurably with new organs, new regenerative processes, a whole battery of life-protecting antibiotics. Pills that show a marked increase in ability to memorize have been tested on animals. Another pill allows a person to get by on only two hours of sleep a day. Pain and pleasure can be controlled by electrical stimultion of the brain. It is even possible to have one animal bear the fetus of another through artificial transplantation. Control of genetic characteristics is part of the experiments being conducted. What may seem fantastic at first glance is rapidly becoming scientific as we find that there are almost no limits to our abilities— we have already gone past the sky, the proverbial limits of aspiration.

Brainstorming is based upon the group approach and is usually more effective that way, but the same procedure holds true for an individual who wishes to explore his own mind. In the group, you are faced with the stimulation of other minds and their capacity for intellectual excitement. The competitiveness of other personalities will help you focus on the problem and serve as a spur for renewed effort. In addition, you can "hitchhike" on others' ideas, coming up with new suggestions that are triggered by their approach.

The group should not be too large because it is harder to "warm up" more than 12 or 15 people, and larger groups tend to become somewhat disorganized as they go in many different directions at once. A moderator or leader is needed in the background to keep things under a minimum of control, to prevent any negative judgments or criticisms, and to write down the ideas as they come out. Keeping a running list is very important as it helps bridge the gaps when everyone seems to have run out of ideas and stops to consolidate thinking. At that time, the leader should go over what has been proposed already to refresh memories and start discussions going again. Almost invariably, new things will be suggested before the list is read to completion. But this process will probably have to be repeated three or four times in the course of an hour or hour and a half session. (Two hours are about the limit for most people before their enthusiasm wears thin.)

Quantity, not quality, is the keynote of a brainstorming session. The more ideas being thrown up for discussion, (even wild ones) the hotter the topic will become and participants will literally shout out their suggestions. With a free-wheeling group that has had some experience working together, you will probably get 45 to 50 suggestions in an hour, coming in starts and spurts. Of these, as many as one-third might be worthwhile for further consideration.

The creative faculties do not operate freely when they are being censored or when a person is afraid of making a fool of himself. Everyone should be drawn into the discussion, prodded if needed to make him feel that his contribution is valued. It is advantageous to "prime" a group with a theoretical problem that has unlimited answers, all of which could be considered as a probable solution. For instance, sketch a situation in which a man is leaving the country, but failed to get to the dock on time, and sees his ship already on the horizon. What is he to do? How can he arrange other transportation? Where would he go for help? Another open-ended idea, this one with commercial possibilities, is a camera mounted on a tripod with a shutter-aperture that would swivel. If this were geared correctly to the film advance, it would be possible to take a 360° picture, bypassing the expense of trying to construct a super wideangle lens with distortion.

Try not to set up a situation in which the participants have to give value judgments because it might lead to personality conflicts, and answers will come out incoherently, more in the form of vague personal philosophies applicable only to certain conditions. If this does happen, restate the problem to break it down into simpler terms that can be attacked with concrete suggestions.

Though most sessions will generate enough enthusiasm to bring out spontaneous ideas, the moderator may want to ask certain questions leading to sought-for improvements, or he may suggest combining two or more ideas to refine them and bring up better ones. The questions should be broadly phrased so that limits will not be unconsciously imposed.

Things to consider: In the use of a product or process, can new uses be found? Can it be modified in any way? What could be substituted for it that would be cheaper or more efficient? What about combining it with something else? Would it be improved by changing the physical characteristics, i.e. form, shape, color, mechanical power, motion, sequence, etc? If a process or sequence of events is involved, could these be changed, reversed, put in different order, blended? Perhaps it needs new ingredients, stronger or lighter materials, synthetics, etc.

A good illustration of this would be a laser beam. This is a new concept with many properties that offer a tremendous variety of possibilities. It is described as a device for producing "coherent light." Its light waves are all identical in length and frequency and all travel in the same direction, hence coherent. Coherent light does not diverge as does light from an ordinary lightbulb or flashlight and thus does not dissipate its energy with distance. Because the light beams are parallel they can be projected much farther and focused and condensed, thereby concentrating their light intensity and heat energy.

How can a laser beam be *adapted* to various other fields?
   To prevent tooth decay. For art work by painting (exposing) film with rays of light. A low energy laser could be used as a child's toy.

In what way can a laser beam be used to *displace* known methods of doing things?
   In place of x-rays for selected applications.
   As a light source for a camera instead of flash bulbs.
   In the automotive field, as a timer or in the ignition system instead of spark plugs.
   In quality control guages. For excavation, to open oysters, and for a cigarette lighter.

What *easier* way can a laser beam be used to do something?
   To analyze chemicals.
   To mix solutions.

Can a laser beam be *combined* with something else?
   Automatic traffic light signal indicator.
   In a fire truck to control traffic lights.

Can a laser beam be used for *greater accuracy?*
   In land measurement.
   In a light source for a gun sight.
   As a spectrographic standard.

Can *reliability* be better accomplished using a laser beam?
   In automatic guidance systems

In what *substitutions* can a laser beam be used?
   For the detonation of explosives.
   As a separator.

How can a laser beam be used to make something *more efficient?*
  As the third stage in rocket propulsion

In what ways can *greater control* be achieved using a laser beam?
  Extreme heat in a small area.
  Engraving.
  Nuclear fusion.
  Generating steam.

Can a laser beam be used for *improved health?*
  Curing a cold.
  Curing strep throat.
  Sterilization of milk.
  Growing hair.
  Drilling teeth.

Where can a laser beam's *speed* be used to advantage?
  In a laser gun.
  In cooking to minimize wasted heat and shrinkage loss.

How can the laser beam principle be *combined* with other principles?
  Wireless transmission of power.
  Communication with space ships.
  Communication between submerged submarines.
  To drill oil wells.

When working with a group, but especially when brainstorming by yourself, try to picture the completed product or desired end in as much detail as possible, setting up the perfect invention, and then working back to see what would be involved. This will help keep things on the track and simplify the problem by breaking it down into individual segments for analysis as to their practicality or usefulness. You may then have to change the end product because an entirely different approach that is much more feasible may come up in the course of the analysis.

Brainstorming has proved its worth in the business field and is now taught as a required course for top level management and research teams in many large companies. It also offers unique and challenging opportunities for the free-lance inventor or group of inventors who want to hire out their services to brainstorm an idea.

"Associated Ideas" has done it on a nominal pay scale and come up with more solutions in an hour than a large testing laboratory was able to conceive in two years. The group averages about four workable solutions out of 35 or 40 ideas (which is roughly the limits of one session). The members do not have inside knowledge of all the fields they are asked to work in, but often the very newness of their approach is an asset in their favor, as they are able to combine it with a working knowledge of physical and mechanical laws.

One unusual challenge was presented by a lumber company that supplied fresh-cut timber blocks for use as pads for pile drivers to insulate the shock between the hammer and the wooden or concrete pile. Because of the intense pressure and heat involved, the blocks would catch fire spontaneously, and the down-time involved for assembling a new pad was very expensive. Among the ideas they came up with that the company found most acceptable were: asbestos powder that would simply pack tighter by the pile driver force and is not combustible; redesigning the face of the pile driver for greater distribution of heat; flameproofing the blocks; a heat-dissipating pad composed of layers of aluminum and asbestos. For other companies they found uses for coffee grounds, a way to separate trash from tea leaves, a cotton fiber picker, mechanical separators, etc.

Here is a wide open example of a basic patent idea that could take years to develop, but would be extremely profitable. It satisfies all the requirements of an ideal invention—the cost is minimal, the profits are extremely high, and it would be habit forming.

The problem is loosely stated on purpose to give maximum opportunity for personal brainstorming and imaginative solutions. See if you can find an answer, or the best approach to follow, as an exercise in stimulating the mind.

Consider the needs of alcoholics, or an even larger class—those people who do not want to drink especially, but have social obligations in situations where most others are drinking and urge them to do likewise. A problem arises in that many people just cannot stand soft drinks all night long—they are not only satiating because of sweetness, but also fattening and not really "socially acceptable." Coffee or tea is a favorite at many alcoholic centers, but are not always available and tend to irritate the system after too many cups.

Here is a definite need of literally millions of people. A socially acceptable drink that is non-filling, pleasant to the taste, hopefully has some lift to it, readily available, has no bad side effects, and would increase the "sense of well-being."

This would be an ideal drink. More than likely, one that would fulfill all requirements could not be manufactured, but something approximating it could be found. The question to be researched is, "What are the basic essentials that would have to be included to make it acceptable?"

This is where creative inventiveness comes in. Just having an idea or a vague awareness of a need is not sufficient. To be productive one must define the limits of the problem as soon as possible in order not to waste valuable time and money inventing "just for the sake of inventing." A preliminary investigation immediately brings out certain facts that need to be considered before trying to decide on the chemical structure of such a drink.

First it should have dextrose or some other energy supply in it that can be rapidly absorbed into the blood stream without having to be digested. It has been found that alcoholics have a low blood sugar level at various periods of the day and feel a need to drink because alcohol is absorbed directly into the blood stream and gives an immediate lift that compensates for the lack of sugar in the blood.

There is oral gratification or stimulus in alcohol, which should be taken into account. Thus the drink should either be carbonated or have some distinctive chemical to stimulate the taste buds in the mouth.

Then, if possible, consider what would be allowed in a drink to act either as a mild stimulant (such as coffee or tea), or a depressant and or tranquilizer, or something that decreases anxiety.

If such a drink were developed, it would need nationwide advertising as an "in" drink, to compete with present beverage promotions appealing to adults, to those who can discriminate, who want to be different, be a member of the "smart set," etc. Alcoholics Anonymous clubs would be the likely place to test market it for acceptance. They would probably be only too glad to promote something so obviously to their advantage.

While this "free association" thinking is an invaluable tool for opening up new approaches or going past an apparent impasse, it will be worth little or nothing unless followed by a razor-sharp criticism. Now is the time to swing to the other extreme and throw out all the "dead wood" in the ideas you have accumulated.

Brainstorming is just an aid in the larger consideration of finding a need and fulfilling it. Unless the need is kept constantly in mind as the eventual goal of all your work, you may be side-tracked into crackpot inventions that serve no purpose except your own amusement. It is an unfortunate fact that only about one out of 100 inventions submitted to the Patent Office each year becomes a large financial success. It is to prevent this disappointment that you are urged to keep "the need in mind" and match up your invention to it as closely and perfectly as possible.

Some readers at this stage already will have a project in mind that they want to begin work on or have started. Now is the time for them to consider not all the good things in their idea, but rather all the bad things—the imperfections and bugs that must be ironed out before the product can be considered commercially acceptable.

Remember Dr. Robichaux's fish scaler? Though he handles orders in the thousands now, it took 15 model changes before he had an item that could be manufactured easily with a maximum profit return. The first scalers were workable, but too crude for public acceptance or mass production. Essentially made of a wedged-shaped head that had a sharp serrated edge on one side and a handle for grasping, the device seemed utterly simple until repeated tests were made. He found that one metal he tried, cast iron, was not strong enough at the juncture of the handle and head and would break off. The handle itself had to be lengthened by an inch over the first model to give sufficient leverage. Later it was found that a metallic scaler, though easier to mold, was too expensive because it had to be trimmed and finished by hand. Finally, a tough, colorful plastic that required no edge trimming was produced, and it has proved highly successful. The next step Dr. Robichaux has in mind is to mechanize the device for even further speed and ease of use.

If Dr. Robichaux were to have a commercial research firm test his scaler, he would rightly expect more than just an analysis of its component parts. A laboratory is valuable because it employs a scientific, disinterested method of rating a product for saleability to the public. It also provides a service even when it points out that weak points in the invention far outweigh advantages. In the long run the inventor will be better off by learning this and abandoning the project before he has spent hundreds of dollars uselessly promoting it.

You can take the same "laboratory" approach, being careful to weigh truthfully the advantages and disadvantages of your "brainchild." If the weak points are too numerous, drop it, and go on to another idea. However, if you think it has potential, try this checklist for improving a model before going to the test market stage.

Cost

This is the basic consideration to which all others are subordinate. The first rule that must be considered is the realization that the proposed product, if manufactured, must be made for about one/fourth of the final retail price; this is necessary to cover handling cost, shipping, salesman's commission, middleman's profits, at least 45 percent retail markup, and royalty rights.

Products expected to sell in nationwide outlets do best when they are priced under $2, or even better at $1.25 or 88¢, which is the approximate limit of "impulse buying."

An invention with the best hope of success from a cost standpoint would be one that eliminates present steps or products altogether, and replaces it in a simple, efficient manner; for instance, irradiating foods with radioactive isotopes that make them last indefinitely with no canning, freezing, or protective cartons necessary.

Market Limiting Factors

Would it appeal to both men and women, or if slanted, have features that can be exploited as being masculine (strong, long lasting), or feminine (solving household chores, keeping skin attractive).

Is the potential large enough to warrant costly tooling up? If the market is too limited or technical, it may be saturated with only a few hundred items.

Could it have broad application in foreign markets, perhaps utilizing techniques other countries don't have?

Could it be test marketed on a limited basis and still show appreciable results for ease in selling potential ideas to manufacturers?

Timing

Are other technologies able to provide needed backup services (heat resisting plating for new airplane design)?

Be adaptable if something newer is developed. (Better plastics for contact lenses.)

Older generations may be tradition bound (distrust new medicines).

Gimmick, seasonal toys are sometimes planned a year in advance, especially Christmas toys endorsed by personalities.

Would inflation, depression, war, etc. be a hindrance or an aid?

Manufacturing Requirements

If mass produced (necessitating a large market), would dies be expensive; is any close tolerance machining necessary?

Is extensive hand labor involved or could it be set up on an assembly line (hand wiring versus printed circuits)?

If chemical, is there any way to save steps in processing or saving money using less refined materials?

Size

Can it be made more compact, folded up, disassembled for shipping? No unneccessary bulkiness or weight.

Can it be made portable, perhaps with a portable power pack? If a food product, could it be dehydrated to save handling and storage costs?

Could it be sold with the possibility of add-on units or with accessories. (Army Ranger outfit for children that adds on guns, headset, holsters, boots, etc; calculating equipment added on to increase capacity and flexibility?)

Safety

No toxic materials or sharp corners in toys.

Adequate safeguards for electrical devices to prevent accidental grounding or shorting out.

No side effects from combinations of simple chemicals (Bleach and ammonia can produce deadly chlorine gas.)

Moving parts covered up.

Simple to operate without complicated instructions.

Will not impair visibility (A green windshield tint will absorb red color from tail lights or signal lights.)

Efficiency

No parts that will break easily or require undue maintenance.

Long lasting (shoes for children that won't wear out).

Does away with intermediate steps. (Cleaning clothes by sound waves eliminates soap, water, and drying.)

Simplest number of moving parts possible for longer wear and better use of energy source.

Can it be mechanized to save time and labor for buyer?

Does it do a job or process quicker, easier, more completely than anything else (lawnmower that floats on air to prevent drag)?

Can it be made multi-purpose or adapted for use with existing products (adapter to make television set an FM radio receiver)?

Basic Materials

Can it be made from what is now waste material (styrofoam insulation).

Would reinforced plastics be better than metal (cheaper and take high-gloss finish easier)?

Takes advantage of new discoveries (laser, sub-miniature electronics, radioactive materials, new optics, fiber lenses, synthetic drugs).

New construction techniques eliminate many conventional materials (nails, bulky insulation, outside painting, high strength glues).

Better strength materials available (alloys, ceramics).

Legality

If an invention is not in the public interest or is potentially dangerous, a patent will be disallowed.

New laws demand innovations (anti-smog devices, auto exhaust controls, magnetic catches for refrigeratos doors, greater auto safety factors).

Drugs and chemical processes take much longer to be approved because of unknown side effects.

There is a need for better filters for cigarettes, better jamming devices for electronic snooping equipment.

Distribution

Can the product be handled by yourself or will a jobber or national company be needed? If so, will it go with related items of companies already established?

Can it be shipped easily without resorting to expensive containers to protect it?

Could it be sold in vending machines (raincoats)?

Design and Appeal to the Senses

Utilitarian, yet pleases the senses as a consumer item.

Would accessories have to be added on to make it more attractive (chrome finish, bright colors for women or children)?

Consider packaging that is neat and/or disposable (charcoal briquettes in an egg-carton package with its own starter fluid).

Helps eliminate pain (laser beam or sound-wave dentist drill for cavities).

If a liquid product, could the container be used for something else later (easy pour jug, plastic squeeze bottle)?

One procedure followed by a commercial laboratory cannot be stressed too strongly or often enough—the matter of keeping detailed, accurate records of all work done on the invention, even when negative results are obtained. Next to an unbreakable claim, this is the strongest protection you can give to your invention. At this stage it is most vulnerable to an infringement suit (though the court action would not come up until the patent was issued). The recording of continual testing while reducing the idea to practice is the only legal basis allowed to show priority rights. It will also serve to show non-abandonment of the project should the patent claim be contested.

Research organizations, whose very livelihood depends upon accurate accounting of their notebooks, drawings, and analysis sheets, take extraordinary precautions to protect their time and effort in a project. They will have one person, or a whole department, charged with the responsibility of maintaining records that are legally defensible. Over the years, weathering court actions and infringement suits, they have reduced necessary paperwork to a science, with little or no room for error. As the work of many people has to be coordinated, research firms go to elaborate lengths not to duplicate effort, waste man hours, or leave gaps in quality control. Their system is much more comprehensive than one an individual would have to follow, but because it is so complete in covering every aspect, it would be well to outline the procedure. The reader may then simplify the method for his own circumstances.

Establishing priority is the keynote of any system. Data are accumulated not only to prove conception by the inventor, but to show reasonable diligence in carrying the original idea forward to completion. This is most simply handled by using consecutively numbered sheets that are dated and witnessed to provide time-noted receipt.

From the time an idea is first conceived, it should be written down, the paper signed and dated by the inventor. This is done even with the employee suggestion box. Memorandum slips are numbered according to any consecutive system, and note is made of the date received. Some companies even insist that at this stage the idea should be witnessed and corroborated by two persons who understand the subject.

If the project is carried forward for additional work, it is assigned to a researcher who will record all data in a bound notebook, utilizing consecutively numbered pages with no erasures, no blank pages, with incorrect entries initialed, and with every page signed and dated by the researcher. This notebook is as valuable as money in the bank and should be protected from any possible damage. Some companies in highly competitive fields will date and witness *each page* and have the corroborators state that they understand the record. Thus each page stands as a separate piece of evidence and can be introduced in court without having to reveal the content of other pages. This is admittedly a time-consuming operations, but in a highly secret project it may be considered necessary.

This raises the question, often given seemingly contradictory answers, of how long an inventor can wait before having his notebook witnessed by a disinterested party. The law says "a reasonable length of time" without putting definite limits on it. "Reasonable" is usually construed to mean a matter of a few weeks or less; a lapse of one or two months might possibly, in a closely contested suit, be too long. But one point is definite, however—the inventor must not be unduly tardy in making his entries. For an inventor to wait months before recording his work would be to invite allegations that his memory as to date of conception could be false, that he was not really pursuing his efforts with diligence, or that errors could have been made in transcribing late entries.

The matter of dating records holds true also for drawings, tracings, production reports, job orders for supplies, etc. Records of changes in drawings show attempt at reduction to practice even when going up a blind alley. Drawings should not be altered nor should coded numbers on individual sheets be altered. Sheets should also be signed and dated by the draftsman.

A research and development report, making an evaluation of findings and determining the scope and purpose of the invention, is similar to a patent claim in many ways. It defines the limits conceived by the research team, and as such should not have value judgments in it that belittle the project. In the same vein, it should not be unecessarily wordy but leave room for further interpretation and additional uses of the invention. A report unduly detailed might mean that another person could duplicate the same process or product with only mimimal changes and claim that his was just as legitimate. It also closes the door for offshoot inventions or further refinements that may come up as new knowledge is gained in the field.

The individual will not have to make this kind of report to himself, of course, but when presenting his case to a patent attorney he will have to make a summation of purpose and intent similar to it. Therefore he should be guided by the same consideration, to reveal in as much detail as necessary the salient points of his invention, but leave it broad in application to make a stronger claim that will cover a larger field. Determining how to balance what should be included or left out of the actual patent claim is the work of the attorney, or the individual if he wishes to submit his own patent application. This, however, will be covered in Chapter V.

If a patent attorney is consulted, certain pertinent data from your notebook will be asked for. In brief outline these will include not only your name, date of invention, and witnesses, but also steps followed in reducing the idea to practice, the scope and purpose of the invention, where it fits in its field, what problems it answers, description of the invention itself, and probably photostats from some pages of your notebook.

All of these elaborate safeguards are upheld by common and statute law to encourage the inventor and give him security from competition until a patent is actually applied for. But unless an application is filed, these laws are inoperable and the invention eventually goes into the public domain. Patents are primarily for the benefit of the public. To provide impetus and recognition to the inventor, he is given exclusive rights on the use of his invention for a period of 17 years, then it goes to the public. If the inventor attempts to hold back his findings for too long a period of time (usually more than one year) he can be considered to have abandoned the project and loses title to it.

The protection afforded an inventor can perhaps best be illustrated by the language of the court itself in considering the case of *Christie v. Seybold: 55 Federal Reporter 69.*

> He who first conceives, and in a mental sense, invents a machine, art or composition of matter, may date his patentable invention back to the

time of conception, if he connects the conception with its reduction to practice by reasonable diligence on his part so that they are substantially one continuous act.

The three elements generally considered as showing sufficient proof of reduction to practice are the notebook (a running commentary on the idea from the time it was first conceived until patented); drawings, or sufficient written explanation and disclosure so that one skilled in the art can understand the operation of the device or process; and the making of a model. A model does not, of necessity, have to be constructed, but the theoretical explanation of the invention must show that the device is operable.

Another element that does not always enter the picture, but may be the most time consuming one, is that of actual testing. This could take two or three years while a person is further refining and improving his invention for greater performance. This has long been held to be valid because it brings a more sophisticated invention to the public for its greater benefit.

But it must be remembered that after an invention is placed in the commercial market or made public through some general circulation medium, only one year is allowed in which to file for patent application.

The "no nonsense" business-like approach in keeping records should, of course, be extended to include all facets of inventing as a serious occupation. As with any other scientific profession, conclusions and projections of future work obligations should be based on facts, not emotion. As the idea is conceived and brought to a finished stage, money (both potential and working capital) is a prime consideration. An invention entails a large outlay in time and expense, and if it doesn't prove itself somewhere along the way, it should be abandoned in favor of a more profitable idea.

For one moment, consider yourself in the position of a business manager charged with overseeing the costs of 100 inventions. Carried through to completion, the patent applications would run conservatively from $500 to $700 apiece, including patent attorney fees. Another $200 may well be spent in testing and building models. As mentioned earlier, it has been estimated that only about four out of 100 inventions bring considerable commercial success. Some applications would be dropped way before filing fees are asked but still, if only 50 were carried through to full patent issuance, this represents a total of $25,000 to $35,000 wasted for all intents and purposes—money that could have been saved if the inventions had been abandoned when they failed to pass the qualifying tests. If this money had been used instead to promote other ideas the law of averages alone (about 20 to 1) would favor a better return for capital investment.

It is relatively easy to adopt a scientific approach to inventing but the secret of success lies in motivation—the gathering of all the forces and power of one's personality and throwing them enthusiastically behind a dream. This is stressed over and over again in widely-read books that seek to explain what makes some men successful. It certainly is not just luck. It is a driving force that makes some people take advantage of every opportunity to better themselves. They have a clear goal in mind and the faith in their own ability to commit themselves to whatever is required to reach it.

Inventing requires creativity more than anything else. It takes a pioneering, adventurous spirit to question the status quo and look for improvements. One must be dissatisfied with things as they are, but be able to envision better plan and have the courage and persistence to promote it. This is the creativity of an artist, for whom everything he comes into contact with is "grist for his mill, an opportunity for greater advancement and riches."

This change in mental attitude does not come overnight. The enthusiasm of a new idea will carry a person along for a while, but to keep the mind stimulated and derive the greatest benefits from his mental energy, certain principles should be followed. A professional inventor uses them either consciously or through well-developed habit patterns. Though they can be stated in many ways, here are some of the fundamental tenets:

1. Invent in the field you most enjoy and your proficiency will invariably be greater. Do not make the mistake of trying to make a million dollars on a scheme that you know nothing about. You will not only spend too much time trying to learn necessary technical details, but might very well become discouraged. It is important to carry an invention through to its logical conclusion, because otherwise you may be tempted to carry on too many projects at once and realize none of them. That is not to say that you have to work exclusively on one idea. Many find it more enjoyable to have two or three projects going (when one gets stale, concentrate on another to refresh the mind) but it is better to limit it to no more than four at a time. Money will come, through experience and persistence, but you must keep at it constantly to insure success. Just as in any other profession or trade, the apprenticeship takes years before one is fully qualified.

2.   Always look for new applications of the things you use or read about: Make it a challenge to think of all the ways you could change things for the better. Keep note of these ideas and jot them down in your bound notebook and have the entry dated and witnessed. It's something you can always refer to when looking for new avenues to explore and you could very well get the jump on someone else by your early date, even though you don't follow it up right away. To keep abreast of new products, read extensively, especially the science and mechanics magazines, popular scientific journals and newsletters, trade publications in your field and related ones for cross-germination of idea.

3.   Profit by your mistakes. Problems will come up, even complete failure for some projects, but try to salvage something positive from it. Use the library to get all the information possible about your invention; sometimes it can have application in an entirely different area than originally intended, and can be turned into a success.

4.   Have the invention evaluated early on by professionals. Use the services and help of other people in developing it and have a patent search made not only to save money in the long run, but to find out what competition you are facing. If it appears that is has to be dropped, have the good sense to do so. But, once again, keep your records intact. The day may come when newer technology or products are developed that will allow your invention to advance, and it will become very valuable.

Co-author Terrence Fenner recently had an eye-opening experience demonstrating the need to keep records for future use. A few years ago he was working on a permanent press process for clothing but the patent was awarded to another company. However, he kept the general chemical concept in mind and after he left his former employer he revived the idea—a process for use by the housewife at home to give permanent press to clothes already purchased—attacking it from a different angle. Now he has a new patent application filed and companies seeking him out that were not even remotely interested five years ago.

III. Taking A Stand

There are two times when the fledgling inventor is most susceptible to being exploited by con artists and patent pirates or misguided by well-meaning but ill-advised friends or business acquaintances. When an invention passes its first hurdles and looks like a sure thing, enthusiasm runs so high that the inventor cannot wait to market it for the benefit of the whole world. Or, conversely, early in the process, if it looks like the invention must overcome impossible hurdles, the inventor becomes so discouraged he is willing to try anything that will take the problems out of his hands. Both of these conditions are liable to occur in the early development stages before he has the level-headed guidance of a patent attorney or has invested considerable time and money developing his idea by accepted, practical research techniques.

Eschewing the slow, sometimes maddening pace of patent office prosecution, the inventor is tempted to look for a cheaper, quicker, easier way, convinced that he knows enough about his product and business practices in general not to be taken in as sucker-bait. Being invention-minded, he has probably seen at one time or another the long list of classified ads for inventors appearing in the popular, non-technical magazines and feels tempted to try their services. Just the sheer repetition of these ads, appearing year after year, seems to give them a kind of trustworthiness by virtue of their permanence. And, besides, what's to be lost; a few pennies in postage, a few dollars in advance research fees? Everything, if a particularly unscrupulous or incompetent party gains the inventor's confidence.

Let us examine the claims of a few of these ads, and then spell out what they really mean.

1. **Tremendous Manufacturers' Lists.** The result of writing off for this list from one of the many companies that advertise similarly was that one company sent back a list derived from the alphabetically arranged index appearing at the front of the Yellow Pages in a telephone book. The list was so ill-chosen that even clubs appeared in the addresses given as possible sources interested in a particular invention.

37

Others use *The Thomas Register of American Manufacturers*, which lists thousands of firms and the products they deal in. But this book can be bought by anyone, or found in most libraries if a person wishes to promote his own invention. Certainly, it is a hit or miss proposition at best, with this come-on costing as much as a hundred dollars or more if you let the company serve as your advertising agent.

   2. Recommended Procedure or Invention Record. This is simply a fancy disclosure form, looking more impressive than a patent from the government and replete with certificates and ribbons, that puts your invention on record on a certain date at the company's office. This can easily be done yourself, as explained in Chapter 1, without revealing all aspects of your invention to a firm that very probably will use the information to pressure you into letting them develop and patent your idea for only a few thousand dollars.

   3. Licensed Patent Engineer. There is no such thing. Only a patent attorney or patent agent is licensed by the U.S. Patent Office, and these names can be secured upon request from the Patent Office. Licensed attorneys pass rigid examinations, cannot advertise, and are bound by law to uphold the practices of their profession. There is no legal recourse against patent brokers, engineers, former examiners, etc., and the inventor uses their services at his own risk.

   4. We Will Sell Your Invention Or Pay Cash Bonus. After advancing a fee of a few hundred dollars, most of which is eaten up by expenses, if the company doesn't sell your invention they will return part of your fee—usually $25 to $50. In effect, you are paying them to give you a "bonus" of about 10% of the money you advanced.

   5. Professional Patent Searches. $6. The disadvantage of this was mentioned before, i.e. it is impossible to make an adequate search for $6. The usual fee is $35 to $50 dollars. Not only is a poor search worse than none because it leads to false hopes, but it is also used as a come-on scheme to praise your idea highly, allowing the company to develop it for as much as your bank account can stand. The end result may be a patent that is so poorly written and executed that it proves worthless.

   6. Financial Assistance Available. There is nothing wrong with financial assistance. A patent attorney will usually help you get it, or may even take on extended payments for his services. What these companies are doing is selling your name to finance companies that operate at high rates of interest, and often give part of the interest fee back to the advertiser in attorney fees and finders fees. Once a promissary note is signed by you, the obligation to pay remains, no matter what the outcome of the invention, and in some cases they can put a lien on your property if payment is not made on time.

7. Free Patent Facts. Nothing more than brochures about inventing and Patent Office procedures that can be learned in much greater detail from books or by writing to the Patent Office for specific information on problems you may be encountering. The word "free" invariably means a follow-up mail campaign to get your signature on a brokerage contract.

8. Patent Applications Prepared, With Development, $200.00. Once again, a case of an unlicensed person advertising as if he were a patent attorney. In reality, the patent papers will be drawn up and sent back to you, and then you have to submit them in your name and follow through by yourself the difficult, extremely complicated prosecution procedure with the Patent Office, trying to overcome their objections as best you can—a highly risky venture.

9. Over 20 years Experience. All Aspects Covered. Some of these companies that advertise in vague generalities make no false claims, but are just as vague in their answers. Letters sent to them get a form letter response that doesn't begin to answer the question asked. They give vague promises, tell about three or four inventions they were successful with, and declare that if you will register with them success is assured.

10. Patent Broker Wants Inventions. Some of these organizations are legitimate, though expensive, but usually deal with an inventor after he has his patent or is in a patent pending stage. Their marketing and advertising techniques can be done by the inventor at a much lower price, and sometimes more to his advantage because of the personal approach. However, the use of legitimate patent brokers and/or marketing yourself will be discussed in detail in Chapter VII on selling your invention.

There is an agency, one of impeccable reputation, that you may want to deal with, but which can cause considerable headaches—and that is the federal government. Mention has already been made of the equitable program for rewarding ideas of employees working for the government. But what of the independent inventor who wants to sell to the military or other defense agency? Unfortunately, the consensus of opinion expressed in recent polls indicates that the government indulges in too much red tape, delay, and (sometimes) indifference. It seems, on the whole, according to the survey, that working with private industry is not only less vexing but also more profitable in the long run.

In fact, the National Inventors Council, which served as a screening agency for the government, has discontinued its publication of *Inventions Wanted by the Armed Forces and Other Government Agencies*. There was much criticism directed against the proposals in this book for being too vague in defining the problems for which answers were sought and, conversely, being too discriminatory when invention ideas were submitted. The impression given in many cases was that the government wanted the invention fully tested, with all the bugs worked out and ready for production. Inventors naturally felt that if they had their ideas fully perfected there was no real reason to ask for government help—at that stage of the game they could just sell it to the highest bidder, whether public or private.

However, inventions are still being accepted by the military, but information on needs should be secured from the particular branch that could utilize the invention, i.e. Army, Navy, or Air Force.

There are still two agencies that should be of interest to inventors. One, the National Aeronautics and Space Administration (NASA) Washington, D.C. 20546, can serve as a licensee for privately owned patents and will give "reasonable compensation" for these rights. NASA has a tremendous R & D program with over 4,000 spin-off invention ideas a month, and can serve as an extremely valuable source of information for those working in the space complex field. Even though NASA may not buy an inventor's idea, the advanced technology available through its office could serve as a breakthrough in the inventor's particular field. Some of this spin-off data is made available only by contract to businesses that buy the service. Further information about regulations concerning this office can be obtained from a booklet entitled *Patent Laws*, available by sending 35 cents to the Superintendent of Documents, Government Printing Office, Washington, D. C.

The other agency that will work closely with the inventor is the Small Business Administration. Because of the many regional offices in principal cities, the inventor can have the opportunity for person-to-person contact with the SBA representative and receive advice on business practices, resources available in the community for market or product testing and, perhaps, financial aid. A more complete explanation of the functions of the SBA will be found in Chapter IV on marketing.

The SBA, having an intimate knowledge of business conditions, can often guide an inventor to local, informal groups that will help in product research and/or development. These groups are on the upswing now, many of them growing out of university-based scientists who want to join their pure theory pursuits with more practical technicians in the business world. Some universities form consortiums among many schools (such as Research Corporation) and pool talents and money to set up testing laboratories for promoting and marketing inventions.

Another highly successful "incubator laboratory" was begun in Philadelphia to provide guidance, space and supporting services to men with an idea and an urge to start their own business. Charging only about $50 per month per man, it receives additional financial assistance from the U. S. Economic Development Administration.

In connection with this laboratory project, Industrial Research, Inc. interviewed 35 entrepreneurs in the Philadelphia area to identify the source of major problems and what techniques were used to solve them. Highlights of the study showed that most of the men were middle-aged, technical or professional people, working for some other company, and did early development work at home (in the kitchen or basement). Almost 60 percent of the people interviewed said they developed a marketable item in less than two years, but some took up to five years to get started in business for themselves.

Finances were a major problem, with about half of them using personal savings as a base, but in most instances this was not enough and recourse had to be made to banks or private loans. About a third of the founders started out with between $1,000 and $10,000 and another third with up to $100,000.

Well defined markets, conservative operation and experienced management were reasons given for companies showing profits in the first years of operation. However, all said a tremendous amount of work and determination were needed, with many putting in 70 hours a week during the company growth years and 55 hours on average afterwards.

Partly due to the interest generated by the Philadelphia "incubator laboratory," other states are beginning similar projects, financed by both government and local agencies. Some of these are designed to become self-supporting in five years by advancing money to the inventor and asking repayment after the company is firmly established.

Private foundations have underwritten inventor's costs for a number of years, the most widely known being Battelle Development Corporation. It is willing to review the invention and, if accepted, carry most of the costs of testing and marketing for a percentage paid back to the Battelle Corporation. While investing in a wide variety of new ideas, Arthur D. Little, Inc. is more strict in requirements. They seek products that will have a large market appeal and can expect to gross over $1 million—but they have the organization to handle such an enterprise. University staff members can draw upon the services of Research Corporation, which secures the patent and shares the profits with the University and the inventor in an arrangement that is decided upon by the institution and personnel employed there.

All of these corporations, government or privately funded, require a query as to the conditions that must be met for submission and a disclosure form to protect themselves and the inventor.

There is another route open for the inventor that requires him to be his own borker or salesman through the mails, but the extra effort can lead to guaranteed royalty rights. This is done by contacting the Technical Service assistance representatives of medium-sized and major companies for help in ironing out kinks in the product. For example, du Pont ran a series of tests for an inventor to determine what kind of plastic would best suit his needs. Tests covered stress, ease of casting, design limitations, alternate products, etc. Another company, queried on a non-slip compound for wet-weather shoe soles, recommended one of its many latex products, and sent along samples showing optimum use, price lists, related articles in journals and text-books on its effectiveness. In this particular case, the price list of the basic product was the deciding factor in discontinuing experimentation along that line because the finished goods would have been too expensive to market.

The rationale behind this cooperation with the inventor is that if the invention proves marketable, the materials that the company produces will be specified for use in the finished product. The inventor is not bound to this arrangement in case he wants to work with another manufacturer, but experience has shown that inventors will rely upon materials worked with in enough cases to warrant the time and trouble a company takes in this public relations venture. Sometimes smaller, more easily diversified companies become prospects for the inventor's item and finally wind up as manufacturer or distributor of the inventor's work.

Another cooperative venture with ever growing possibilities for inventors is the Inventors' Exposition. Beginning over 10 years ago in isolated parts of the country, these congresses are now held in almost every state, with an international exhibit conducted in New York that attracts hundreds of entries and provides a market place for millions of dollars in royalty and product sales.

The expositions held primarily for inventors are usually jointly sponsored by state agencies, universities, and/or local Chambers of Commerce. Their main objective is to provide an economical means to display inventions to manufacturers, distributors, investors, and also the general public. The latter, a hard to please audience, helps the inventor realize problems that must be faced in perfecting the product for wider acceptance.

Sometimes country fairs, home shows, new retail products, conventions, etc. offer space' to inventors for display. This can be helpful as a certain number of people are attracted to the exhibit, but unless the inventor is already in the production stage, he will not be meeting the right people for his particular needs. Even if he has a finished product, criticism is made that the fees (often as high as $100 to $300 for booth space) could be better spent in other advertising or promotional ventures.

The Inventors' Congress aims at a different stage in product development. Manufacturers' representatives and others who come can visualize the inherent possibilities in an invention even in early stages. And that is when they want to negotiate for production, licensing, etc., before the invention is completely perfected and committed to a company for manufacturing. About the only advantage in bringing a finished product to a congress of this type is if you are looking for additional distribution outlets, better sub-contract arrangements (if applicable) or, perhaps, financing if you are in business for yourself and are considering expanding. But these goals are secondary to the main business of helping inventors get exposed to prospective buyers and receive technical help, as most congresses have experts in the fields of licensing, patenting, processing on hand to give group lectures and private audiences.

A typical state-sponsored exposition will last for three days and cost about $25 for booth space. The fee usually covers the cost of at least one banquet and entitles the inventor to attend the lectures given by specialists in various fields of interest to the inventor. Arrangements are often made to have affairs of interest for women too, as inventors often bring their wives along, especially when they have to travel to another city for the meeting. A wife can be an asset in manning the booth as someone should be there at all times to explain the invention. Often it is hard, or impossible, to single out a prospective buyer from the general public. Sometimes, these company representatives do not want to commit themselves until they have gone back to their company for approval for further negotiations, and thus will not reveal themselves fully.

The inventor should also consider whether he will have extra costs for signs or special construction. Because of this, he should make arrangements as early as possible with a congress official so he will know exactly what to plan for in terms of money, food and sleeping accommodations, shipping of exhibits, etc.

The better established congresses seem to have many of these problems well worked out for the convenience of inventors, but some of them do not give enough advance notice, are short on attracting manufacturer's representatives and do not engender enough publicity to bring in large numbers of the general public. However, as associations grow to promote this exposition service, and receive better backing from inventors themselves, many of the problems will be solved and these congresses will grow in importance.

Special mention should be made of Patent Exhibits, Inc., which holds a yearly international exhibition in New York City. This fall congress is by far the largest and best managed of them all   It attracts businessmen from all over the country, thousands of spectators, and deals in millions of dollars worth of sales. But it is expensive. Booth space for the 1966 convention ran $350 for 10 days, $250 for four days. Standard display items, membership in the association, listing in the catalogue, booth signs, are covered by the initial fee, but any extra construction or services requested must be carried by the inventor.

Dates of this exhibition, as well as those that are state-sponsored, are available at no cost from the Office of Inventions and Innovation, National Bureau of Standards, Washington, D. C. 20234.

This office has prepared a guideline for inventors to increase the effectiveness of their exhibit, and help them derive greater benefit from the congress. It has been derived from the observations and experiences of a number of inventors and should prove of value in deciding whether or not one wants to take advantage of this type of exposition. The following points should be noted:

1. Make every effort to register several months in advance of the exposition date. Early registration assures you of display space and permits the sponsors of the exposition to plan the event more efficiently and effectively. It also gives the sponsor much needed information on which to base publicity releases used to attract businessmen to the exposition. An advance publicity release on your invention might be

instrumental in attracting the very type of patron you hope to meet at the exposition.

2. See that your invention is covered by an issued patent or a pending patent application. The exposition sponsors cannot be responsible in any way for the premature disclosure of ideas, designs, trade secrets or technical information pertaining to the exhibits. The exhibitor must assume full responsibility for disclosure and any negotiations resulting from disclosure. Remember, too, that your patent attorney can be helpful by reviewing any proposed contract and seeing that appropriate safeguards are incorporated in it for your protection.

3. Have an attractively painted working model, if possible, and be prepared to discuss in detail the features of your invention that you believe will excite the general public.

4. Make sure your exhibit includes a lettered sign with the title of your invention and your name and address (sometimes this is provided by the sponsors). It is also helpful to have printed information on your invention that a company representative can take back and submit with his report to his employer.

5. See that your exhibit is in place and ready for display when the exposition opens and keep it manned until the closing hour. There is nothing more disconcerting to a businessman than to come to an exposition and find only a portion of the inventors there. His time is valuable and he may not return or support future expositions if inventors do not demonstrate the required enthusiasm for accomplishing the desired objectives.

6. Use the exposition as a means of obtaining consumer reaction to your invention. Ask the visitors what they like about it, what they don't like about it, and if they would buy it if it were available on the market.

7. Do your homework and try to be realistic about the significance and potential impact of your invention. Try to visualize the steps and the risks that will have to be taken before you or anybody else will be able to achieve some measure of commercial success. Nobody can fully predict the pitfalls and opportunities, but if you can equip yourself with some feeling for both, negotiations with potential sponsors will be more meaningful and rewarding. Remember, you are championing an idea and that requires a mountain of hope and confidence on the one hand and tactful reflection on the other. Put yourself in the other person's shoes and then ask yourself, "What's in it for me? Why should I be interested in this invention?"

IV. Test Marketing

Undoubtedly one of the highlights of this book is the joining together of the inventor and Junior Achievement* to test market his product.

Junior Achievement is an exciting example of citizenship, of applied economics and business statesmanship. Young adults, 15 to 19 years old, increase their business knowledge by actually controlling the destinies of miniature companies.

Each company, composed of approximately 15 teenagers, is assisted by adult businessmen. These men, acting as advisers, are representatives of a local business concern, civic or professional service group. They are experts in the field of production, business and sales.

The teenagers first decide upon the type of company they wish to operate. They raise money to finance these companies by selling stock at $1 a share.

Although Junior Achievement, through financial support of local business firms, provides the business center, office and production equipment, each teenage company pays a nominal sum for rent, capital deposit, and other necessary expenses. The company also pays its members wages and commission. Junior Achievement has often been referred to by educators as a "super-curricular" activity for high school students.

In the process of running their own companies, teeage members learn to keep various production records, plan sales campaigns, design advertising and prepare financial statements. In short, from October till May of each year they go through every step of owning and operating their own business.

Junior Achievement serves as a necessary supplement for regular school work and provides a "learn-by-doing" laboratory for the students economic education.

Not all inventions are of a nature that will lend themselves to a

* The authors wish to thank Junior Achievement and the U. S. Small Business Administration for permission to edit and publish material used in this chapter.

Junior Achievement test marketing program; hence, the following are recommended additional aids used successfully in test marketing.

Ask your friends' opinion

Your friends are fertile ground for evaluating the worth of your new product. Ask them to evaluate carefully and to try to aid you in improving it.

Make use of trade shows

Sample items on sale at a show that caters to the market you most desire to reach will give a rapid evaluation of net worth of your product. The time and effort to hand-make (if necessary) your item will be well worth the investment. You might consider sharing a booth with a fellow inventor, as Dr. Robichaux and Ralph Carboni did with their products.

Seek store buyers' opinions

Discuss your invention with the people who will buy the invention for stores. It is these buyers' business to know what is likely to sell.

These sources usually will state their reactions to, and opinions on, the following phases of new product development and selling:

Possible uses.

Product salability.

Who will buy the product.

Necessary design changes.

Size, color, style, packaging.

What markets should be promoted most actively.

How product compares with similar items in price and features.

Get them to comment on your price (possibly it might be too low). See what constructive criticism they can offer.

Try mail order

Advertise your product in the mail order shopping section of a magazine that caters to the market you want to reach. (House and Garden, Boy's Life, Popular Photography, etc.)

The average cost of a one column three inch ad is $250-750.

Offer your product to a mail order house like Sunset House or Spenser's and if they refuse your offer, write again and ask them to comment why they were not interested and explain to them that this information will be valuable to you.

Do a one-store evaluation

Concentrate your efforts on one store and thoroughly evaluate why the customers did or did not buy your product. Station yourself near the item to see people's reactions. Did they even notice your product? This information will be of great value to determine the future plans for the product.

Use your product as a premium

Your product in many cases could be test marketed along with a well-known product as an incentive purchase for the latter in competition with its competitors. Enclose a coupon to determine interest in reordering.

Use free publicity

A well written letter to a large number of magazine editors. an attractive eye-catching photograph of your product and a concise news release will often obtain thousands of dollars of free publicity for your product. Visit your local library to obtain the names and addresses of the magazine editors. If possible, send your letter, photograph and news release to every editor of every magazine listed in *Standard Rate and Data* as you never know which ones would be likely to give you coverage.

Write to companies

This method, believed to be dangerous by some because it may possibly give companies the idea for an invention, may be used to obtain an opinion of the commercial merit of the invention. Write select companies and ask if an invention designed to perform a certain function (which your invention does) would be of any value. The positive replies may be followed up with the formal introduction of your invention.

Try direct mail to the consumer

Many products may lend themselves to direct market testing by the consumer. A well written letter, a brochure describing the product, a sales order form and a return reply envelope comprise the basics for this direct approach. If the item is geared to a specific market, contact list brokers who could furnish you select market listings, usually on sticker labels. The cost is from $10 per thousand upward depending to the market selected. In New York contact the following mailing list brokers:

Names Unlimited, Inc.
   352 Fourth Avenue
   New York 10020, New York
James E. True Associates
   419 Fourth Avenue
   New York 10010, New York

Many others may be found in *Standard Rate and Data.* A telephone or street directory is helpful for market testing a general interest item.

Television

Try to get coverage on a variety show to demonstrate your new product or send the station a sample for them to give away as a prize. If possible, make arrangements with the TV station for a movie of your part of the program. You can use this to good advantage in an automatic projector. Dr. Robichaux has an excellent movie of his appearance on Associated Ideas Television program which was a five minute spot on the Bob and Jan Carr "Second Cup" program.

Laboratory testing

The services of the country's commercial testing and consulting laboratories are used by thousands of manufacturers to solve many tough, new-product, material, processing, and manufacturing problems. Their services are also belpful in testing new products before release to the market. These laboratories put a product through grueling tests and usually the "bugs" are eliminated before customers find reasons to complain. The tests will also show how a product compares with the products of leading competitors, in what way it is superior, or If inferior, how it must be improved to meet competition.

Despite the frequent success of product tests by friends, such tests cannot always be relied upon. For example, a manufacturer of a new line of tools had tested them in his plant but uncovered no serious defects. However, as soon as the tools reached the market they began breaking. The company began getting many returns. A commercial laboratory found that the trouble was caused by the method of heat-treating the steel used in the tools. Testing in a commercial laboratory often includes consumer tests under actual-use conditions and, in this case, might have located the flaw sooner.

Commercial laboratories may be located through the following directories: *Directory of Commercial and College Testing laboratories*, American Society for Testing Materials, and *Industrial Research Laboratories in the United States*, National Research Council.

Usually, the price for laboratory testing is not prohibitive and the results are invaluable.

Why test market

Clearly a test marketing drive is not inexpensive. However, it is the only way a company can obtain really reliable information about its product, about the impact it will make on consumers and about the degree of demand it can continue to expect. Moreover, the test makes it possible to develop a formula for the most efficient way to introduce the product in other areas. The further an invention is developed, the more valuable and salable it becomes.

It is well at the outset to recognize that test marketing can mean different things to an inventor than to a large, highly departmentalized business. The latter firms, with long experience in introducing new products, have actually reduced the test-marketing operation to a fairly precise science. The inventor can afford neither the money, time, nor manpower to undertake test marketing on that scale. He must therefore employ it on a basis within his means and cababilities.

Purposes

For both the large and the small, however, the purposes of a test-marketing operation are identical. Before investing in broad sales, advertising, and promotional programs, the businessman will find it profitable to engage in a test to determine:

The marketability of his new product under reasonably normal sales conditions.

The effectiveness of various sales appeals.

The resistance to the new product at various price levels.

The response to different advertising and promotional plans. These objectives apply both to consumer and nonconsumer products. The techniques of test marketing these two types of products vary, however.

Consumer Products

Selecting a test market

It has proven to be a fallacy to select a test market simply because it has been highly successful for other products. The characteristics of your own new product should in a very large measure determine your choice of a test market. Columbus, Ohio, for example, might be an excellent test market for many food products, but not the most ideal for a suntan lotion; Richmond, Virginia, might be excellent for the testing of some sports goods, but hardly ideal for ski wax.

In selecting a test market, the small business owner would do well to find one that fulfills the following requirements:

1. It should be a city or area reasonably close to his home office so it can be supervised at minimum expense.

2. It should be one that is large enough to have advertising media capable of getting his new product message to a large segment of potential users at minimum cost, yet not so large a market that the expense of doing this will be prohibitive.

3. It should be one having characteristics that approach reasonably closely the average of his national market as he envisions it.

In addition to meeting those objectives, it should be possible to determine in the test market a sales, advertising, and promotional campaign that will be applicable, with only minor modifications, to other markets nationally, both large and small. It therefore behooves the inventor to select a test market with a diversified population, income, cultural levels, disposable income, and so on, approximating the national average. It would be best if the market were relatively self-contained, isolated from other cities, had a balance of industry and agriculture, a level of business activity that does not fluctuate throughout the year, and channels of distribution for which the new product is suitable.

Launching the test operation

In kicking off a test operation, it would be well to approach it with the attitude that it is a national program, giving it neither more nor less attention and effort than one would a larger project. A promotional kit explaining the new product and the support it is being given in the way of advertising, promotion and publicity should be made available to the men making wholesale and retail calls.

Duration

The length of time necessary for a test will vary according to the new product's normal cycles. A cereal, dessert, shaving cream or similar fast-turnover item might require only a 13-week program, with a 5-week saturation effort. Slower moving products such as shoe polish would need a longer time for the housewife to purchase, use and reorder them, and have this pattern reflected in dealer reorders.

Cost

The cost of a test operation is naturally subject to considerable fluctuation. A single example will highlight this fact: One manufacturer allocated a budget of $100,000 for a Syracuse, New York program, while an inventor known to the authors conducted a test for $100. In any case, about 60 percent of the expenditure can normally be expected to be allocated for advertising and about 40 percent for sales, handling of cash-redeemable coupons, merchandising, and related functions. Larger operations usually result in a 50-50 division.

Design for Selling

Nowadays a successful product not only has to be good, it has to "seem" good to the customer. There was a time when the attitude of many manufacturers was: "We make our product the best we know how, and we sell it as cheaply as we can afford to. People can buy it if they want to." This better mousetrap theory of merchandising worked as long as there was not too much competition.

But in our competitive world the customer is boss. Whether the product is a breakfast food or a machine tool, it has to sell itself to the customer. The best product is not just best in some abstract sense— it is the one that makes the consumer lay down money and say "I'll take that one."

Appearance is not Everything

The object of design for selling is to make every product its own salesman. Anything that makes a product do a better selling job is interesting to the expert on design for selling. Often when people think of making a product attractive the first question that comes to mind is "Can we make it look better?" The first industrial designers—specialists in design for selling—were called in by manufacturers who had an idea that products would sell better if they looked better. Industrial design became known as a technique for making products and packages beautiful.

Here are three suggestions that you would do well to bear in mind.

First, keep things simple. Except for some articles of fashion like women's dresses and hats and cosmetic packages, few products gain by looking complicated.

Second, don't use disguises. Although the tendency to make things pretend to be what they are not is dying away, it still takes an effort to avoid fanciness and false fronts.

Third, note that in appearance a very small change may make a considerable difference. Extreme and violent changes in appearance are rarely wise unless they signify a real change in the structure or function of the product. In fact, the problem in many fields is to strike a satisfactory balance between giving the consumer the impression that something substantially new is being offered and retaining a recognizable character that has been impressed on the public mind over a period of years. A manufacturer of alarm clocks and his designer solved this problem by working out for several years in advance a series of small appearance changes that followed manufacturing improvements, giving the product a cleaner appearance and at the same time not sacrificing the recognition factor that the original appearance of the product had established in the public mind.

Color

Of all the elements in appearance, color is often the most influential. In recent years, advances in various branches of chemistry and the technology of finishes have made more color available. As a result, the public has become more sophisticated and demanding.

Color as a factor in design for selling is often a direct link between questions of taste and questions of manufacturing technique. If fire-engine red is chosen for a certain part on an appliance, to give it liveliness and zip, and also to make it an eye catcher on dealers' shelves, the next question is—how shall this part be finished? All sorts of considerations are involved in the choice between a baked finish, a quick-drying lacquer, or other kinds of surfacing, or perhaps making the part in a plastic of this color and thus eliminating finishing operations. Thus, while bright red handles on a toaster or hair dryer might be good attention getters in appliance stores, they would be too obtrusive in a breakfast nook or on a dressing table.

Color as a factor in merchandising is tricky and full of unpredictable and unexplainable quirks and inconsistencies. Just as no one has ever explained why housewives in one city will pay a little more for white eggs and those of another area will pay extra for brown ones, there is no rational explanation for the fact that the same color in a line of earthenware will sell in one part of the country and not in another. Sometimes a color will go in the metropolitan bargain basements but not in the higher priced stores. In addition, there are lasting regional preferences and there are passing fads.

Color is influenced by fashion. A color or shade that becomes "the thing" in the fashion world may become an influence in a great variety of fields. A popular textile color will influence the colors available in bathroom tiles, for instance. On the other hand, a color that becomes widely used in architecture is sure to have its influence on textiles.

Despite the importance of fashion and fad in the world of color, it would be wrong to assume that color is an entirely irrational factor. It is a subject that is being studied constantly—by psychologists, by chemists, by market researchers, and by many others. There are well-established facts concerning the effects of various colors on the human being. If color has or could be made to have an important effect on the sale of your product, you should put your choice of colors on as scientific and factual a basis as possible. For specific guidance as to sources of information on color write to the Inter Society Color Council * an organization in which a wide variety of professional and research groups are represented.

Do not neglect the "color conditioning" of the interior of your plant if you decide to manufacture your invention. Psychologically correct color treatment of working space is a major element in creating pleasant and efficient working conditions. If you are not well skilled in the choice of colors, it will pay to get help from a color consultant, an architect, or a designer.

How to Name the New Product

In advertising and selling a product, the brand name might be as important as a good product or a good package. If the manufacturer decides that a brand name is desirable, the final selection should be made

*Box 155, Benjamin Franklin Station, Washington 4, D. C.)

with full appreciation of its value, particularly as an identifying factor in the brand preference of the consumer. In choosing names for products, a manufacturer may (1) coin a name (*Kodak*); (2) adapt and adopt words (*Keen Kutter, Perfection*), or (3) use a name under license or agreement (*Hopalong Cassidy*).

A successful brand name, whether coined or adopted, has a number of general characteristics that can be used as a guide in naming the product. However, a manufacturer who elects to use a name by agreement considers it worthwhile to disregard some of those points in favor of what he considers other advantages. To the extent possible a good brand name is

Short   Pronounceable in only one way
Simple   Always timely (does not get out
                                         of date)
Easy to spell   Adaptable to packaging or label-
                                         ing requirements
Easy to read   Available for use (not in use by
                                         another firm)
Easy to recognize   Pronounceable in all languages
                                         (for goods to be exported)
Easy to remember   Not offensive, obscene, or
                                         negative.
Pleasing when read
Not disagreeable sounding
Easy to pronounce

Package Design

Elements of a good package

In designing the package for a new product, certain objectives should be kept in mind. A good package should

Protect the product
Protect the customer
Carry the product in convenient quantities
Help sale of other products in the line
Advertise and stimulate purchase of the product
Help the sales force to sell the product

Keep marketing costs down
Help reduce the amount of returned goods
Provide necessary information to the buyer.

We can learn from the youth in Junior Achievement who use the forms below and pattern our anticipated market testing expenses accordingly.

Determination of Capital Needs

Normally it is the second month before sales income is received, so *initial capital* should be based on costs for at least the first two months. If sales are anticipated earlier, or later, your figures should be adjusted accordingly.

Below is a suggested form for figuring initial capital.

Estimate

 1. Wages
 2. Salaries
 3. Raw materials for two months
 4. Fees
 5. Rent for two months
 6. Company records, materials and forms
 7. Rental of furniture and fixtures
 8. Medical payments—insurance
 9. Rental lease of equipment
10. Advertising and promotion
11. Postage and office supplies
12. Bad debts
13. Banking and service charges
14. Total

$

How to Price a New Product

Pricing new products is important in two ways: It affects the amount of the product that will be sold and it determines the amount of revenue that will be received for a given quantity of sales. If you set your price too high you will likely make too few sales to permit you to cover your overhead. If you set your price too low, you may not be able to cover out-of-pocket costs.

What is Different About New Products?

New products that are original require a different pricing treatment from old ones because they are distinctive. No one else sells quite the same thing. This distinctiveness is usually only temporary, however. As your product catches on, competitors will try to take away your market by bringing out imitative substitutes. The speed with which your product loses its uniqueness will depend on a number of factors. Among them are the total sales potential, the investment required for rivals to manufacture and distribute the product, the strength of patent protection, and the alertness and power of competitors.

Although competitive imitation is almost inevitable, the company that introduces a new product can use price as a means of slowing the speed with which competitive proucts are placed on the market. Finding the "right" price is not easy, however. New products are hard to price correctly. This is true both because past experience is no sure guide as to how the market will react to any given price, and because competitive products already on the market are usually significantly different in nature or quality. Therefore, in setting a price on a new product you will want to have three objetives in mind:

1.   Getting the product accepted.
2.   Maintaining your market in the face of growing competition.
3.   Producing profits.

Your pricing policy cannot be said to be successful unless you can achieve all three of these objectives.

What Are Your Choices as to Policy

Broadly speaking, the strategy in pricing a new product comes down to a choice between (1) "skimming" pricing and (2) "penetration" pricing. There are a number of intermediate positions, of course, but the issues are clearer when the two extremes are compared.

Skimming pricing

Some products represent a drastic departure from accepted ways of performing a service or filling a demand. For these, a strategy of high prices, coupled with large promotional expenditure in the early stages of market development (and lower prices at later stages), has frequently proven successful. This is known as a skimming price policy. There are four main reasons why this policy is attractive for new and highly distinctive products:

1.   The quantity of the product that you can sell is likely to be less affected by price in the early stages than it will be when the product

is "full grown" and competitive imitation has had time to take effect. These early stages form the period when pure salesmanship, rather than price, can have the greatest influence on sales.

2.   A skimming price policy takes the "cream of the trade" at a high price before attempting to penetrate the more price-sensitive sections of the market.   This means that you can make more sales to buyers who are willing to pay high prices for a product they want, and at the same time build up experience useful later in hitting the larger mass markets with tempting prices.

3.   You can use price skimming as a way to feel out the demand. It is frequently fairly easy to start out with a high price that some customers may refuse, and reduce it later on when the facts of the product demand make themselves known.   But it is often difficult to set a low price initially and then boost the price to cover unforeseen costs or to capitalize on a popular product.

4.   High prices will frequently produce a greater dollar volume of sales in the early stages of market development than will a policy of low initial prices.   When this is the case, skimming pricing will provide you with funds for financing expansion into the larger volume sectors of your market.

Penetration pricing

Nevertheless, a skimming-price policy isn't always the answer to your problem.   Although high initial prices may safeguard profits during the early stages of product introduction they may also prevent quick sales to the many buyers upon whom you must rely to give you a mass market.   The alternative is to use low prices as an entering wedge to get into mass markets early.   This is known as penetration pricing. This approach is likely to be desirable under the following conditions:

1.   When the quantity of product sold is highly sensitive to price, even in the early stages of introduction.

2.   When you can achieve substantial economies in unit cost and effectiveness of manufacturing and distributing the product by operating at large volume.

3.   When your product is faced by threats of strong potential competition, very soon after introduction.

4.   When there is no "elite" market, i.e. a class of buyers who are willing to pay a higher price in order to obtain the latest and best.

While the decision to price so as to penetrate a broad market can be made at any stage in the product's life cycle, you should be sure to examine this pricing strategy before your new product is marketed at all. This possibility should certainly be explored as soon as your product has established an elite market. Sometimes a product can be rescued from a premature death by adoption of a penetration price policy after the cream of the market has been skimmed.

The ease and speed with which competitors can bring out substitute products is probably the most important single consideration in your choice between skimming and penetration pricing at the time you introduce your new product. For products whose market potential looks big, a policy of low initial prices makes sense, because the big multiple-product manufacturers are attracted by mass markets. However, if you set your price low enough to begin with, your large competitor may not feel it worth his while to make a big production and distribution investment for slim profit margins. For this reason, low initial prices are often termed "stay-out" prices. In any event, you should appraise your particular competitive situation very carefully for each new product before you decide on your basic pricing strategy.

Factors You Should Analyze in Setting a Price

Once you have decided on your basic pricing strategy, you can then turn to the task of putting a dollars-and-cents price tag on your new product. In order to do this, you should analyze at least five important factors:

1. Potential and probable demand for your product.
2. Cost of making and selling the product.
3. Market targets.
4. Promotional strategy.
5. Suitable channels of distribution.

How Junior Achievement Prices a Product

In order to get a retail price for a J. A. company product, a number of considerations are taken into account. With the help of the information that follows and the break-even chart below, you will be able to arrive at a satisfactory price. You will note as you try different figures that the more units you produce, the lower the cost per unit. Here you have one principle of our economy, which is based on mass production and broad-purchasing power.

Estimating costs and margin (see break-even chart)

1. Estimate total number of units to be produced in your program
2. Estimate overhead and administration costs:
   Rent of facilities and machinery.
   Shop and office supplies.
   Wages.
   Packaging and promotion.
   Administrative & miscellaneous (charter fee, bank services, insurance, etc.)
3. Estimate total raw materials costs.
4. Allow for commissions, contingencies, etc.
5. Allow for profit
6. Total costs, expenses, and profit margin:
7. Divide (6) by number of units to be produced to get approximate retail price.
8. Add sales tax if required.
9. Suggested retail price including sales tax.

New product costs may be segregated into half a dozen main categories:

1. Direct labor.
2. Materials and supplies for production.
3. Components purchased outside.
4. Special equipment (jigs, dies, fixtures, and other tools).
5. Plant overhead.
6. Sales expenses.

Direct labor

Methods of estimating direct labor may be built up in one of three ways: (1) Comparison of each operation on each component with accumulated historical data, from your files, on similar operations for similar components. (2) Development of a mockup of the proposed workplace layout, and actually timing an operator who performs the series of manufacturing operations, simulated as accurately as possible. (3) Application of one of several systems of predetermined, basic-motion times currently available from private sources. Make certain, however, that you include any added time used for setup work or needed to take the item from its transportation container, perform the operations, and return the item again to its transportation container. When the total direct labor time is determined, multiply it by the

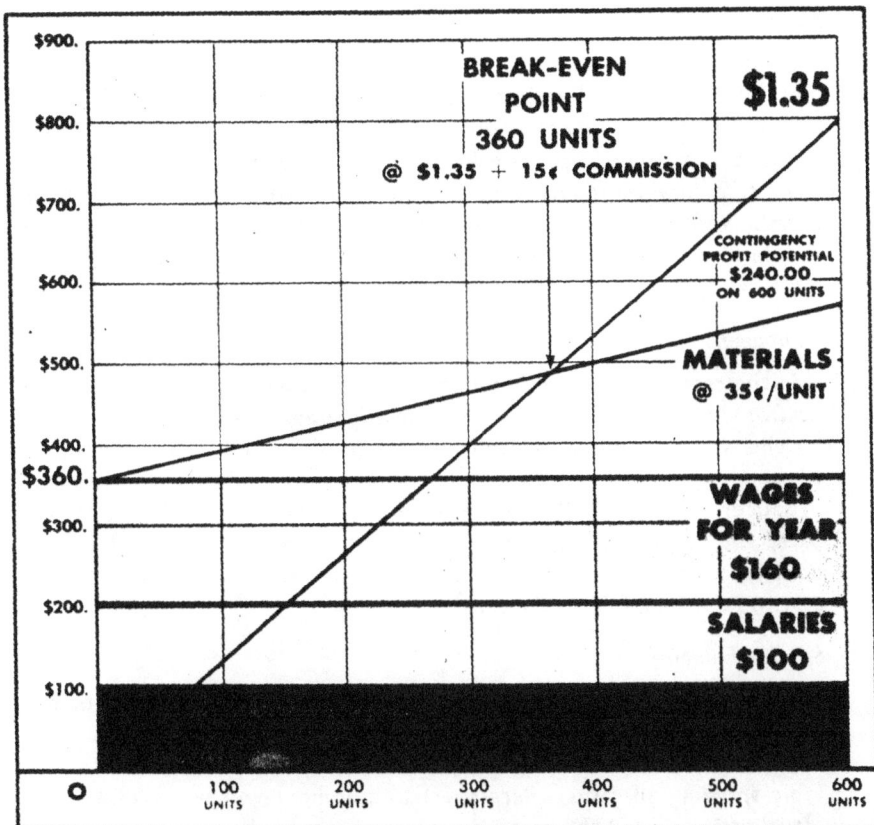

BREAK-EVEN POINT 360 UNITS
@ $1.35 + 15¢ COMMISSION

$1.35

CONTINGENCY PROFIT POTENTIAL $240.00 ON 600 UNITS

MATERIALS @ 35¢/UNIT

WAGES FOR YEAR $160

SALARIES $100

A Break-Even Chart enables you to determine how many products your J. A. company must sell before it shows a profit. Construct your own chart at the start of the year to help in deciding on a price for your product. Use the fixed costs for Wages, Salaries and Overhead shown in the sample. Add to this base your materials cost. Then test possible prices (first deducting any tax and 10% for commissions) to see what sales goal is needed to yield a reasonable profit.

Later in the year, you can work out your position more accurately, since you'll have a better idea of fixed and material costs.

In the sample shown, a price of $1.35, plus 15¢ commission and sales tax, was selected by a company expecting to produce 600 units. The breakeven point is 360 units, or $540.00. A lower price would have lowered sales income and cut profit to an unsafe margin. A higher price would increase possible sales volume, but could be so high as to create customer resistance and bring about a year-end inventory of unsold goods.

Figure 3.   Break-even chart.

appropriate labor rates.

Materials and supplies for production

In developing reliable cost figures for materials and supplies, make a methodical list of all requirements. Having listed everything in an organized fashion you can enter the specifications and costs on a manufactured-component estimate form. Remember to include any extra costs that may be incurred as a result of requirements for particular length, widths, qualities, or degrees of finish. Allowances for scrap should also be made as accurately as possible and corrected by applying a salvage factor if the scrap can be sold or reused.

Components purchased outside

In the case of parts purchased from other concerns, place your specifications with more than one reliable supplier. Get competitive bids for the work. But in addition to price considerations, be sure you give proper weight to the reputation and qualifications of each potential producer. Moreover, if you use a substantial volume of purchased parts you may want to use a 'plus'' factor above the cost of the components themselves to cover your own expenses involved in receiving, storing and handling the items.

Special Equipment

Take careful precautions against making a faulty analysis of your expense and investment in special jigs, dies, fixtures, and other tools needed to produce the new product. To avoid trouble in this area, make a table showing all cases where special equipment will be needed. The actual estimating of the costs of such equipment is best done by a qualified tool shop. Here again, competitive bidding is an excellent protection on price. Do not include costs of routine inspection, service, and repair; these are properly charged to plant overhead.

Plant overhead

The overhead item may be estimated as a given percentage of direct labor, machine utilization, or some other factor determined by your accountants to be the most sensible basis. In this way you can allocate satisfactorily charges for administration and supervision, for occupancy, and for indirect service related to producing the new product. Overhead allocations may be set up for a department, a production center, or even (in some cases) a particular machine. In calculating plant overhead make certain that in setting up your cost controls your accountants have not overlooked any proper indirect, special charges that will have to be incurred because of the new product.

Sales expense

As in the previous cost categories, the critical element is the *added* sales expense that the new product will involve. To make sure you have included everything, it is often helpful to deal with these expenses in several segments. The following is a simplified checklist: (1) salaries, commissions, and traveling expenses; (2) advertising and sales promotion; (3) transportation; (4) credit and collection expenses; (5) warehousing and storage; (6) sales overhead expenses—including office expenses, insurance, depreciation, and the like. Other lists could, of course, be developed, but for greatest usefulness they should be kept as simple as possible and should be organized in terms of the specific activities.

How To Tell When a New Product Is Successful

Keeping records of repeat orders.

One hardware manufacturer, who has successfully introduced a number of new products in recent years, maintains that he can tell within a matter of months the relative degree of success or failure of his new item by keeping an accurate record of the repeat orders that result from its introduction. This experience is also echoed by sales managers who have successfully launched fast-turnover food and drug products. The analysis of such data will indicate quite readily the degree of trade and consumer acceptance. In this respect it is well to study the chart depicting the cyclical trend followed by new products during the introductory period.

Warranty control

A successful appliance manufacturer has gaged the success of his new product over many years by a combination of factors: Every new item he sells carries with it a warranty, and for the warranty to become effective the ultimate user must fill in and mail back to the manufacturer a postpaid registration card with his name and address, the date on which the appliance was bought and the name and address of the retailer from whom it was purchased. At 6-month intervals, the manufacturer communicates directly with many of the users of his new product to determine their reaction to it, and whether or not they would recommend it to their friends.

Service calls

Some manufacturers of products that require initial installation and subsequent maintenance and repairs state that their field service organization, which has direct contact with the ultimate user, is a most reliable method of evaluating their new product's success and potential. The methods used by various manufacturers of this type of product range all the way from detailed indoctrination of service men down to a periodic casual spot check.

User research

For both consumer and nonconsumer items, one of the commonest methods of evaluation is to select a representative number of ultimate users and have either your own personnel or an independent research organization interview them. In the case of a nonconsumer or industrial product, relatively few would normally be required. In addition to determining the acceptability of the product in this way, the manufacturer also can learn the following:

Is the product being purchased as a result of advertising, professional recommendation, publicity, or for some other reason?

Is the ultimate user fully aware of all the attributes of the new product?

Is the product versatile in its use, and does it accomplish the purposes for which the user bought and applied it?

Will the purchaser buy this product again or recommend it to others?

What are the shortcomings of the product, the limitations of its use, the restrictions on its application or performance?

What is the reaction to the price the user had to pay, and what is the user's opinion of its value in terms of the price?

Trade research

A representative number of interviews of people who have bought the product can elicit the following significant data to contribute to judgment as to the success or failure of the new product:

The adequacy of retail and wholesale discounts in sustaining trade interest and cooperation.

The degree of confidence in the trade that the new product can achieve sufficient sales volume to warrant continued time and attention.

The effectiveness of countermeasures against competition.

The effect on the trade of the advertising, merchandising, and promotional plans.

The need for price maintenance in the face of competitive tactics.

Where To Go Now

After a successful market testing comes the big question: Shall I manufacture the product or try to sell it to a company? The following discussion may aid you in your decision.

Repeated studies show that new products fail more often because of ineffective or inadequate marketing than for lack of merit in the product itself. Hence, the emphasis here is on an appraisal of the many marketing factors that must receive your consideration in launching a new product.

Basic decisions

One of the basic decisions to be made is how to market the new item geographically. Should it first be sold locally, regionally or nationally? In the case of nonconsumer goods suitable for several types of businesses or industries, these questions take the form of "should we sell one to industry, then another, or should we sell to all suitable industries simultaneously?"

How big is the total market?

The size of the total market should not be a matter of conjecture. It should be determined by a careful statistical appraisal of the potential for all products similar to yours. This potential should be analyzed by types of consumers or industries. Determine how many potential customers there are. To get the clearest possible picture for marketing, distribution and sales purposes, try to have this data broken down —geographically, by industry, by price group, by unit and dollar sales potential. The United States Department of Commerce, trade associations, trade publications, and large national weekly and monthly general magazines frequently have much data already available and provide it on a complimentary basis. The Small Business Administration's publications provide a wealth of material. Your own or an independent research man or organization can supplement this information but at a cost.

Is the total market growing?

Obviously, a new product in a burgeoning new industry would have a better potential than a new product in an industry that is in eclipse and declining. Even latecomers to the frozen orange juice, ball-point pen, detergent, synthetic fiber, transistor, television and other dynamic "new product" fields have built substantial profitable businesses even when the point of difference between their product and competitive products has been minor. It takes rare courage to enter an industry that, statistically, is declining. And a manufacturer launching a new product for which there promises to be only a temporary market, as in the case of a seasonal fad, must recognize the short-range potential involved and evaluate his risk accordingly.

How much can you expect to capture?

Will the amount you can sell represent sufficient volume to undertake all the effort and expense necessary to launch the new product? A common mistake made by enterprising manufacturers is falling under the spell of an "iffy" survey—"if" he can sell one family in a hundred, he reasons he can do well. This can be faulty and dangerous conjecture. It may be that the true potential is *one family in a thousand*—a percentage of the entire market so small as to make the new product economically unsound.

What are the competitive products, prices, and marketing policies?

Exactly what will your new product be up against in terms of competitive performance by other products, price disadvantages, special trade discounts, advertising allowances? It is well to recognize that competitors will not make it easy for you to get business and you must anticipate every tactic they are likely to employ to resist your introductory efforts. Historically, it has always proved to take more in money, time and effort to get a new item established than it has to hold a market position for an entrenched product.

What will be your pricing policy?

The pricing policy you announce in connection with your new product should also include a definition of your policies with respect to advertising allowances, credit and collection, the sale of goods f.o.b. factory or delivered, your policy on returned goods, cancellation of orders, and credit for damaged or unsatisfactory merchandise.

Legal considerations

1. Is the product patentable? If a formula is involved, is it properly protected?

It is unwise to take anything for granted with respect to a patent. This aspect of the new product should be placed in the hands of a competent attorney and his assistance employed to be certain that it does not infringe on any previously patented article and is, in turn, protected from possible infringement in the future.

2. Have legal steps been taken to protect the new product, the new brand name and trademark in the United States and abroad?

It may be wise to consider doing this early in the planning stage even though the firm may not contemplate doing business in other countries for many years to come. Generally, it is not considered necessary to do this in all foreign countries, but only the principal ones having an interest in that particular product.

3. Are there any state laws restricting the sale of this product?

In rare cases, some states have laws that are restrictive on out-of-state products that compete with dominant local industries. It is well to have this checked.

4. Are there any state laws affecting the manner in which the new product can be advertised or displayed?

Many states have laws that specify what may and may not be shown in advertising and display material, particularly with regard to alcoholic beverages. An attorney can advise you regarding laws dealing with privacy and copyright, appropriation of ideas, standard release forms, testimonials and prohibitions against lotteries, flags, and money reproduction, and restrictions on premiums and coupons.

5. Has an attorney examined all the initial advertising and promotional material?

He should also examine and approve the text on packages and labels, sales agreements and all literature pertaining to the new product to be certain it does not violate any state or federal laws.

6. Are you familiar with the Unfair Practices Code as it applies to the field being entered with the new product?

If not, it is suggested that you write to the Federal Trade Commission for advice and assistance in voluntarily complying with FTC-administered laws. In this way one can check to be certain that the proposed method of operation does not violate any of its rules. A copy of the Federal Trade Commission Act may be secured by writing to the Superintendent of Documents, U. S. Government Printing Office, Washington 25, D. C.

7.   If the new product is likely to be sold from door to door, are you aware that certain cities have ordinances applying to the activities of such canvassers?

These ordinances can disrupt a plan seriously if one proceeds without knowledge of them.   An attorney, advertising agency or trade association can assist in securing the list of cities so affected.

8.   Are you aware that premiums and certain types of contests in connection with the sale of merchandise are specifically prohibited by law in a number of states?

The launching of a new product very often is accompanied by an intensive premium campaign or a contest.   Detailed information on the legal restrictions as they pertain to both types of activity should be secured.   In any event, the material regarding such premiums or contests, if not kept out of the states affected, must at least be reviewed by an attorney to see that it carries a specific disclaimer regarding such areas.

9.   Is the product going to be sold by mail?

If so, check with the postal authorities to be certain that the product can be mailed (some products, such as matches, cannot, except in certain specified containers), that the method of selling by mail, the claims made for the product by mail, and so on, do not violate established postal regulations.

The foregoing observations highlighting the legal considerations in launching a new product involve policy decision on your part.   Any one of them, and others not mentioned, could prove to be a boobytrap that could damage a new product venture.   For this reason it is suggested that an attorney be made party to all plans from their inception, despite the fact that this might seem to be an unnecessary precaution and an added burden during a period when it is trying to lauch a new product.

Financing the new product

Many authoritative surveys pertaining to the mortality of new products place "inadequate financing" high on the list of reasons for failure. The small business owner undertaking a new product venture is often faced with the loss of the time, money and effort involved in launching the new product.   In undertaking the new product he must recognize that there is always a strong possibility that it will fail.   Answers to the following questions may assist the small business owner in arriving at such decisions.

1. Is working capital adequate during the first year the product is on the market?

If products are successful, the business is immediately faced with ever-increasing requirements for working capital (largely accounts receivable and inventories).

2. Have you anticipated the need for new permanent capital as the new product develops volume?

In the event a new product catches the public fancy even faster than anticipated and necessitates investments in land, buildings, machinery and equipment far beyond present requirements, what does management propose to do about securing the additional permanent capital this will require?

3. How long will it take the new product to reach a break-even basis? Except in the case of long-margin specialty items, few new products are expected to reach the break-even volume within the first year or two.

4. Realistically, what is the long-term profit outlook for the new product?

In the final analysis, is the new product worth the time, money and effort in terms of your long-range corporate plans?

V. Patents

Up to a certain stage the inventor has to contend only with problems of his own making—getting the bug sout of a model, testing, checking the "state of the arts" to prove uniqueness of invention. Now, however, he must face an "ogre" not of his making but one he cannot get around—the Patent Office.

To the uninitiated, it may seem that the Patent Examiners are playing a game of cat and mouse with his application. A patent application is expensive, time consuming, and discouraging. The inventor will run up against what seems like a creaky, grinding, impersonal machine that devours a patent application and sends it down myriad lost avenues of bureaucratic red tape.

Complaints of inefficiency and undue delay in handling patents have been voiced almost since the first patent was issued in 1790 (a chemical improvement for potash). Since then more than $3\frac{1}{4}$ million patents have been issued, and complaints continue. Even Congress investigated the Patent Office. However, the investigating subcommittee found the system basically sound and, indeed, is the product of a remarkable bureau that does a precise, efficient job—not only to protect the inventor and his rights, but also functioning as a vast storehouse of scientific information available to anyone who needs it.

It is because of the care that the Patent Office takes in processing an application that it seems overly cautious, almost as if it wanted to refuse an application. But this, in reality, operates for the inventor's protection because when he does receive a patent issuance he can be sure that every check was made to see that the patent grant was justified and valid.

The main stumbling block in the system, however, is its vast size. $3\frac{1}{4}$ million U. S. patents and over 6 million foreign patents have been issued and the rate of new applications, which grows every year. In

This chapter was prepared with the aid of registered patent attorneys Mr. Calvin J. Laiche, Trombatore, Vonderstein and Laiche, Kenner, Louisiana, and Mr. C. Emmett Pugh, Drury, Lake and Pugh, New Orleans, Louisiana.

1965, 67,000 patents were issued and 90,000 applications received. In addition, the Patent Office handles requests for 25,000 copies of patents a day. The Patent Office has in the past taken up to three years or more to issue a patent because of the tremendous backlog of work. Now reduced to about two and a half years, it is hoped that with new techniques a further reduction to 18 months will be possible by 1970. Also, all patents on file are being microfilmed to provide quicker access. This is extremely important to scientists and inventors because the Patent Office contains more practical commercial technological information than is available from any other source. Many patents discuss the difficulties associated with previous research, development, and productive techniques and offer specific methods for overcoming such problems. This not only gives clues as to how to solve present problems, but serves as a springboard for new innovations and techniques. Often patents contain valuable information not published in any periodical or book.

Because of the complexity of the patent system, it would be well for the reader to familiarize himself thoroughly with the terms used by the Patent Office so that there will be no confusion when making requests for a precise type of action or retrieval of information. The glossary following, while not as extensive as that which a patent attorney must be familiar with, contains most terms that an inventor needs to know when involved with a patent application.

Glossary of Patent Terms and Useful Information

Abandonment (expressed or implied)—The failure to "continuously" act to bring an invention from an idea state, to a reduction to practice, to filing of a patent application.

Allowance—Used by the Patent Office to indicate that a patent will be granted on an invention.

Annual Indices of Inventors and Assignees—The Patent Office indexes patents each year by the name of inventor and by the name of the assignee of record at the time of issue. This is printed as an Index of Patents. Commencing with the volume for 1955, the Index also contains a list of patent numbers issued during the year, arranged in sequence by the class and subclass number in which the patents were classified at the time of issue. Copies of the index may be found in public libraries of the larger cities. These annual index volumes and the weekly issues of the Official Gazette subsequent to December 31, 1962, can be used to supplement the information contained on the microfilm lists in bringing a patent search down to date in a library.

Appeal—To take a patent case to a higher authority.

Assignment—Transfer of the interest in an invention either partially or entirely to another person or company.

Board of Patent Interferences—A section of the Patent Office delegated to hear testimony from two or more parties claiming title to an invention and to determine the original and true inventor.

Chemical Patent—A patent covering either or both a new composition of matter or a new process to produce a composition of matter.

Classification Bulletins—Bulletins supplementing the Manual of Classification by providing definitions and search notes describing and illustrating the kinds of information or patents that can be found in the individual classes and subclasses. Copies of the bulletins are sold by the Patent Office at prices (minimum 10¢) based on their size. A list of bulletins currently available can be obtained from the Patent Office on request.

Claims—That part of a patent application and specification in which the new features of an invention are described and for which patent rights protecting the invention are obtained.

Composition of Matter—A patent covering a new material itself. For example, an invention resulting from a mixture of two or more ingredients producing a new substance with different properties.

Declaration—A part of a patent application in which the submitter or applicant declares that he believes himself to be the original inventor and that his invention fulfills the statutory requirements for a patent. Can be filed in lieu of an oath.

Design Patent—A patent on an original ornamental design for an article of manufacture.

Design Patent Classification—Separate design (patents for ornamental designs) classification covers 93 main classes, each identified with a title and a number preceded by a "D" (to distinguish them from the other classification), and the main classes are divided into subclasses for the more specific aspects of the design, e.g.

CLASS D1, ADVERTISING

(Subclass) 3, Card racks and receivers.

Electrical Patent—A patent on an electrical device or electrical mode of operation.

File Wrapper Search—A collection of all the correspondence between an inventor and the Patent Office. This is available to the public, at cost, upon request after the patent is issued. A file wrapper search of the closest reference cited against an invention would be beneficial in possibly overcoming the rejection.

Filing Fee—A fee charged to file a patent application with the Patent Office. (See pp. 95-96 for exact prices.)

Foreign Patents—Patents issued in countries other than the United States of America.

Infringement—Violation of the rights to a patented invention.

Interference—When two or more patent applications conflict by reason of claim to the same invention, or copying the claims thereof in a pending application.

Invoking an Interference—Deliberately filing a patent application having the exact claims as that of an issued patent to determine the rightful ownership with respect to the first person to conceive and reduce an idea to practice.

Lead Line—A line in technical drawing connecting a name, number, or letter to a particular part of an invention.

Manual of Classification—printed copies of schedules that give the class and subclass numbers and titles (but not patent numbers) together with an alphabetical index of subject matter.

Mechanical Patent—A patent for a mechanical device, or an original combination of mechanical parts. Unlike a process patent, the object itself is patentable, not the method of its production.

Microfilm Lists of Patents—The Patent Office has prepared on microfilm lists of the numbers of the patents in each of its subclasses issued on or before December 31, 1962, and many libraries have purchased copies of these lists.

Model—A working device covering features of the invention.

New, Useful and Novel—the necessary features in legal terms applied to an invention.

Newly Issued U. S. Patents Available on Microfilm—The complete specification and drawings of all newly issued U. S. patents will be available on 16 mm. microfilm beginning with the first patent issued in January 1966. This new service is being offered by the Clearinghouse for Federal Scientific and Technical Information in cooperation with the Patent Office.

Oath—A part of a patent application in which the submitter states under oath that he believes himself to be the original inventor. A declaration in lieu of an oath may now be filed.

Official Gazette—The weekly publication of the Patent Office is the official journal relating to patents and trademarks. It contains a selected figure of the drawing and one claim or description of each patent granted during the week, and helpful indices.

Original and Cross-Referenced Patents: Subclass, Lists—Where a patent discloses subject matter provided for in two or more subclasses, a copy is placed in each subclass. To distinguish these copies from each other, the copy placed in a subclass selected as the primary basis of classification is called an "original" and all others are called "cross-references." For each subclass separate of lists patent numbers are maintained for "original" and "cross-referenced" patents. These lists give the patent numbers and the classification only, not the patentee's name or date of issue. The number of patents in a subclass and the number of sheets on which they are listed are both subject to change due both to patents issued each week and to reclassification.

Patent—An exclusive statutory right granted to an inventor by the U. S. Patent Office to exclude others from making, using, or selling the invention during the seventeen year term for which the patent is granted. The patent itself may be bought or sold either wholly or in part.

Patent Agent—A skilled person who by training and experience can handle and prosecute an invention before the Patent Office. He is a non-law graduate and cannot practice before any court outside of the Patent Office. Only registered patent agents are allowed to practice before the Patent Office.

Patent Application—A set of papers consisting of an oath or declaration, petition, filing fee, and a specification wherein the invention is described and the claims presented.

Patent Applied For or Patent Pending—Terms both meaning that a patent application has been filed with the Patent Office on a particular invention.

Patent Attorney—A graduate lawyer skilled by training and experience to handle and prosecute an invention before the Patent Office. Only registered patent attorneys are allowed to practice before the Patent Office.

Patent Fees—Monies paid to the Patent Office for filing, transactions during prosecution, and issuance of a patent. An exact fee schedule is published in the Appendix.

Patent Office—an agency of the Department of Commerce located in Washington, D. C. having the authority to grant patents giving exclusive ownership to an invention for a period of seventeen years.

Petition—A part of the patent application, formally asking the Commissioner of Patents that a patent should be granted on an invention.

Plant Patent—A patent on an asexually produced new and distinct variety of plant (other than a tuber propagated plant).

Power of Attorney—The placing of all authority to act in one's behalf in the hands of an attorney.

Prior Art—General knowledge of the features of an invention known to those skilled in the art, mainly from information contained in issued and expired patents.

Prosecution—The legal procedures necessary to obtaining a patent.

Public Knowledge—General knowledge of the features of an invention known to the public.

Process Patent—A patent on any new and useful process or method to produce a product.

Public Search Room—The Patent Office maintains a Search Room at its office in Crystal City, Va. (suburb of Washington, D. C., in which all patents are arranged in groups according to the class and subclass titles of the Manual of Classification. This enables any member of the public to browse or search all patents in any field of invention and to obtain copies of those that he finds to be most interesting. It is ordinarily advisable to engage a patent attorney or agent to make a search before a decision is made to file a patent application.

Reduction to Practice—To perform successfully the features of the invention, e.g. by making a model, running a chemical reaction, etc.

Reference Numeral—A number in a patent drawing used to label a particular part of an invention.

References Cited Against a Patent Application—Issued patents, publications, or anything of public knowledge to act as a statutory bar against the issuance of a patent covering the same public knowledge.

Rejection of a Patent Application—Refusal of the Patent Office to grant a patent for good and sufficient reasons, e.g. if the invention is not new or is obvious.

Searching—Examining the patents and publications on record with the Patent Office to determine the newness or patentability of an invention, the validity of an issued patent, or the existence of infringement.

Specification—A part of a patent application in which the features of the invention are described and the claims presented.

Statutory Bar—Legal reasons why the Patent Office will not issue a patent. For example, an over-one-year-old public use, sales or printed publication describing the invention whether authorized by the applicant or not; the applicant is not the original inventor, etc.

Subclass Subscription—By prepayment of a deposit and a service charge, an inventor may have sent to him, as they are issued, future patents classified in the subclasses containing subject matter in which he is interested. For the cost of such subscription service, a separate inquiry should be sent to the Patent Office.

System of Patent Classification—The Patent Office has arranged its patent collection in groups divided by subject matter. There are over 300 main groupings, called "Classes", and each of these is identified by a number and a title (e.g. Class 2, Apparel). Each of these classes still contains too many patent copies for ready searching, so they are further divided into numerous smaller groupings called "Subclasses," also identified by a number and by a title setting forth specific details or features of the patents in the subclass, e.g.

CLASS 2, APPAREL

(Subclass) 144, Neckties

There are more than 64,000 subclasses and these form the basic units into which the patent classification system is organized. While the classification printed on any patent is correct at the time of the patent issue, it should be noted that the constantly expanding arts often require reclassification. As a result, the classification indicated on the patent may be incorrect at a later date.

Before a formal patent application is filed, it is quite often feasible to conduct a prior art search. This can be done either by yourself, a professional search organization, or through a patent attorney. Determining the best time for conducting the search is up to the individual. There are advantages and disadvantages to both an early or a late search. Basically, you want to first determine if the idea has already been patented. With so many patents in existence, chances are that there is probably some work closely related to the invention you are working on. If so, an early search will show what improvements have to be made to warrant the rating of your invention as "new and novel." Also, this research on the "state of the prior art" will help sharpen your thinking and point up ways to refine your model or process. It may even indicate that the field is overcrowded with inventions of your kind, and thus its chances of being patented and successfully marketed would be limited.

Additionally, an early search is recommended for an invention that requires a great deal of testing and development. Otherwise, a large investment in time and money may be made only to find later that the finished product has already been patented. Early searching orients one in the field and indicates how his patent claims will have to be dawn up to circumvent the claims of related patents.

The main drawback to an early search is that the idea may not be sufficiently developed to allow accurate comparison and cross-check with related patents, with the consequence that some of these patents would be inadvertently overlooked.

A preliminary patentability or pre-ex search generally only gives a feel for the "state of the art." It generally cannot be guaranteed to be 100 percent perfect because it is seldom economically feasible to conduct a "fool-proof" search. Rather, it gives reasonable assurance that your invention, or some portion of it, is novel enough to warrant pursuit of a patent application.

There are basically three approaches that can be followed in making a preliminary search: do-it-yourself; employ the cut-rate, classified searcher; or employ a professional search organization, patent attorney, or patent agent (who may use their services also, especially if his office is outside of Washington, D. C.)

Many popular science or mechanical magazines carry advertisements by various individuals offering to make a search for as little as $6-10, and sometimes offering to process the patent application itself for an inventor. The reader should be warned about this type of service. First, duly registered patent attorneys and agents are never allowed to advertise. However, there is no control over those who do advertise and they may take your money (perhaps hundreds of dollars if carrying through the whole application) without giving you proper representation, as a result of which you may fail to secure any patent coverage. Even if qualified, they are not professionally bound to obtain the best patent coverage for the applicant. The net result is that they may take or obtain weak patent coverage just for the purpose of obtaining a patent for you. In such a case, it is better not to have a patent since your claims will probably be easy to circumvent and you will find that you have merely dedicated your broad contribution or ıdea to the public.

The sheer magnitude of 300 classes and over 60,000 subclasses of the Patent Office makes it financially impossible for anyone to give a legitimate search for under $10. Just the time involved in the search room would mean that the person being paid to do it was not getting enough money and consequently could not economically devote sufficient time to each search. In addition, there is the office overhead of advertising costs, correspondence costs, price of patent copies, mailings, etc. To get a bad search is quite often worse than obtaining no search at all, because the inventor would proceed to spend a much greater amount of money for his patent application, only to find out that his idea had already been patented and was not allowable.

As brought out above, it is possible to conduct a search yourself, even though it is time-consuming, especially if you live outside of Washington, D. C., or do not have access to a library having copies of patents on file or on microfilm. However, it can be accomplished if the library near you has copies of the U. S. Patent Office Official Gazette and files of issued patents by number or on microfilm.

Here is a suggested procedure to follow when making your own search by mail:

Write the Commissioner of Patents, Patent Office, U. S. Department of Commerce, Washington, D. C. 20025, explaining your invention in as much detail as possible (include a drawing or rough sketch if it is required for an appreciation of the invention), and request the class and subclass in which your invention would be classified. It is important to be specific by pinpointing your contribution, otherwise the Patent Office may send you many extraneous lists of classes and subclasses or miss your patentable idea entirely. Naturally, they are not responsible if your search is faulty and a later search by them during prosecution of your application turns up an invention that duplicates yours. However, they are competent, and if your description is sufficiently accurate, they will send all pertinent information.

After you have received the above information, you then order the subclass lists of patent numbers (at 50¢ per sheet of 100 numbers) and cross reference patent numbers. You can then roughly check the subject matter of these patents by reading the claim(s) published in the Official Gazette. You will probably have to check a list of 300 or 400 patents. Out of this you will probably find six to 10 patents which seem close enough to yours to warrant ordering copies from the Patent Office (cost 50¢ per patent).

The above system is obviously time-consuming. However, it does offer the advantage that it will thoroughly acquaint the searcher with the "state of the art" in his field of interest. If you cannot understand the oftentimes technical jargon of a patent, it might be unwise to rely upon this method of search.

The best method from a time standpoint, as well as for accuracy, is to employ the services of a professional searcher in Washington. He can physically go to the search room located and maintained in the U. S. Patent Office for this purpose and go through every patent and all related ones to conduct his search. He also has the advantage of being able to get help from attendants in the search room who have been working there for many years. They quite often know not only where to find the answer more quickly, but can also refer the searcher to a related field that may shed additional information of use to the inventor. A competent professional searcher is an expert in his field.

Since it is impossible to know about all fields, a searching organization may divide its work up amongst its members who specialize. Consequently, such organizations are not only more efficient, but often have information in their files that is not in the Patent Office. For instance, large companies often publish their own technical journals, which reveal new directions in a field and describe latest techniques. They also use this medium to make patentable ideas public knowledge. Thus IBM might invent certain pulley arrangements in one of its machines that the company doesn't want to take the time and trouble to patent. By publishing it, it becomes public knowledge, and no one else can get a patent on it after one year of date of publication. Even a Dick Tracy example (such as a wrist watch radio) may be filed by an examiner to be cited against a patent application.

Patent searches generally run from $35-50 for a relatively simple invention, $100 and up for a more complex search in many related fields, e.g. an electronic device using many components. A patent attorney or agent usually does not make any additkonal charge for this service. His fees in relation to a search cover his time in analyzing the invention, handling correspondence, and advising the client how to modify or improve his invention in light of the information received through a search. (A list of search organizations in Washington, D. C. and patent attorneys and agents can be found on p. 306.)

Before submitting a patent application or contacting an attorney to do so, the inventor should get some idea of the financial risks involved. A patent in itself is seldom a guarantee of success, whether employed in your own business or for receiving royalties when sold to another. A patent is but one building block in a business enterprise. The real foundation of the business rests upon the acumen and aggressiveness of the people engaged in it. A patent is primarily a kind of insurance against competition, but not 100 percent assurance that someone else will not try to enter the same market with a similar product and take your customers away. Ideas are a dime a dozen and it takes a certain kind of knowledgeable person to turn them into profit. An invention is only the starting point in a long process, and many inventions never reach the marketable stage.

Because of the time and money risks involved, an inventor should evaluate his situation carefully to decide when to apply for a patent and how far to pursue it. This decision is fairly clear-cut in the case of a "basic" invention (e.g. first phonograph, incandescent light, laser beam). If there are no basic flaws in the invention and it represents a truly new and practical approach to a problem that may open up a whole field of endeavor, then by all means patent it with claims as broad as possible to retain exclusive rights to the whole field. The chance of realizing a return on investment in the near future is slight since a whole range of subsidiary devices may have to be perfected to make the original invention commercially feasible. On the other hand, the long-range chance of high profits will outweigh this disadvantage. A good example of this is the magnetic door lock for refrigerators. This was slow in getting started, but now all new refrigerators must have it by law to protect children from climbing into an abandoned refrigerator or freezer and being locked in when the door shuts.

Some manufacturers, however, will hold back a patent on a basic invention until they can develop it sufficiently to make sure they corner the whole market. Xerox Corporation did this as well as Polaroid Land Corporation, which has over 70 patents in the U. S. on its film alone. It has sometimes happened that a basic patent was so far ahead of its time that it never reached the market until after the 17-year protection period had expired. With technology moving ahead so fast now, the chances of this occurring are steadily diminishing.

Most inventions are improvements upon existing products and represent almost a reversal of procedure in determining the risks to be taken.

An unfortunately large percentage of inventions that are mere improvements of construction of a technique do not warrant the expensive tooling up procedures that a business has to make for large quantity

production. Because of this it is necessary to take a calculated risk in
market testing the invention and holding it back from immediate
patent pending application until it has proved its worth. This can be
done by submitting it to a selected list of manufacturers for their approval and recommendations, or by producing the item yourself and
selling it on a limited basis to determine public acceptance. In both
cases, however, remember the one-year limitation once it becomes
public knowledge. Also, a disclosure form must be completed between
the inventor and manufacturer to protect the inventor.

Most manufacturers are honest and do not want unfavorable publicity that would come in a court fight by their attempt to steal an idea.
Also, they know that if the invention has been submitted to more than
one manufacturer, the other one may fight them if the idea has commercial merit. The remote possibility is there, however, and is a risk
that must be considered. Some manufacturers who want to use an
invention will assume the financial cost of getting it patented because
they want the patent to be as strong as possible and be tailored-made
to their particular situation.

If you have market tested an idea or used some method to determine
its commercial acceptability, and have made a preliminary search and
found that apparently it is new to the field, you may want to apply for
a patent and then submit it to business concerns while the patent application is pending. At this point you have a clear presentation to
make to the manufacturer outlining just what your invention com-
prises.

But this may also act as a limitation because once the patent is filed
there can be no basic changes in the application, i.e., no new subject
matter can be added. Thus, a manufacturer can draft a patent that gets
around your basic idea and submit it himself and avoid the necessity of
paying you royalties. This can happen if the claims are very specific
and can be circumvented by minor alterations. It is the responsibility
of the party drafting the application to particularly avoid writing weak
claims. However, it is not possible in all cases to protect fully some
inventions, especially when it is a narrow or "specific construction"
type of improvement.

Once you have patented an invention and set yourself up in business
or have sold rights to another concern to manufacture and distribute
your invention, there is still another hurdle that may arise—namely,
infringement. Generally, the chances of infringement are remote;
however, it does occur in about 1 out of every 100 cases. But it can

happen, especially on an invention that is a minor improvement over existing ones in the field. As a practical matter, because of the considerable expense of infringement suits, the parties settle out of court for an amicable royalty fee. The courts, as well, grant leniency to the person who infringes upon another patent without specific intent, i.e. knowing that that person acted in good faith in making a search and pursing a patent through proper channels.

Where economy is a consideration, a patent attorney may advise his client not to try to patent a marginal improvement invention. A good patent attorney can get a patent on just about anything if he fights hard enough and especially if he is willing to take weak claims just to get the patent. However, since the patent would cover a very limited area, perhaps the person would be better off not getting a patent at all. If a person wants to manufacture and sell the invention himself and a search shows that it does not "definitely" infringe existing patents, it might very well be cheaper in the long run to go ahead and get the business started, and if an infringement suit does evolve, then he can settle out of court for a percentage of the profits. The person would at least have the business going and would have saved $1000 to $3000 initially. Yet again, a patent should be viewed as a business expense that has certain advantages as "liability insurance" that would be of benefit as in the above situation.

It should be noted that the infringement of a patent gives the patentee the right to enjoin an infringer from further infringement as well as secure damages for past infringement. In effect the business could be shut down.

It is possible to get a patent on an invention and still not be able to use it. This can happen when another patent "dominates" it, i.e. has broader claims that cover the specific improvement of the "lesser" patent—for instance, if you have a chemical process that would combine reactant A with reactant B in the presence of a catalyst at a temperature of 350°F to give a 99 percent yield of compound AB with no catalyst loss. In searching the prior art, you find that the closest patent to your process gives only a 90 percent yield because it does not use the precise temperature and pressure ranges (covers, for instance, a 200-400° range) and employs the same catalyst with some catalyst loss. Your invention is obviously superior and represents a material gain in knowledge that under most circumstances would be sufficient to get a patent for your improvement over the broader process. However, in a subsequent infringement suit, the owner of the "broader dominant" patent might very well be able to prevent your using the superior process without payment of royalties. Not only that, with the new knowledge gained he could easily employ your teachings and it would be very hard to prove that he was doing so; however, he technically would be infringing your patent.

In the mechanical field, suppose you develop a substantial improvement in a valve lifter that would only be good for use in one make of automobile and would not be adaptable to other cars. You could go to all of the expense of getting a patent on this; however, if the car manufacturer did not want to buy it, then there would be nothing you could do. Without his cooperation, your patent would be for all intents and purposes useless.

These unusual cases are not brought up just to discourage the inventor, but merely to point out that not only must a thorough search of prior patents be made, but more important, the commercial potential of an invention must be investigated.

Anyone can make a patent search and/or file a patent application for another party. However, only the inventor can execute the oath unless legally incapacitated or dead.

In the latter instance, the administrator or executor of the estate or guardian may apply. Two or more persons who make an invention jointly apply for the patent as joint inventors. Mere contributions of finances or technical help (such as in perfecting a model) does not give a person the right to file as inventor. Rights to an invention may be assigned at any time, but it is still in the inventor's name.

An inventor may choose to have a patent attorney or patent agent handle his case. When he does so, all documents that he would ordinarily submit himself are then submitted by the attorney or agent for him with the representative having a "power of attorney" in all matters relating to the patent. Patent Office action and replies will be sent to the attorney with a duplicate copy made available for the inventor's own files.

It is important to choose an attorney or agent with care. A list of registered attorneys or agents can be bought from the patent office for $1. Most large cities have them listed in the Yellow Pages of telephone directories. However, the reader is advised that personal contact with a representative close at hand is best because many problems can arise for first-time inventors that are difficult to answer only by mail. The services of agents (who are fully qualified to handle patent proceedings before the U. S. Patent Office) generally cost about 20 percent less than those of attorneys. However, only a patent attorney can plead a case and appear in a party's behalf should litigation arise. In addition, contractual agreements between the inventor and a company he may wish to assign his invention rights to have to be drawn up by an attorney. If an attorney has to come into a patent case after considerable action has been taken, he will necessarily have to make additional charges for "reading up on the case" and getting background to defend the client. For a complicated and important invention, it would more likely be cheaper in the long run to employ a patent attorney from the very beginning.

Attorney's fees are normally based upon the amount of time he has to actually spend in preparing and processing an invention over the normal two to three year period it may take to get final issuance. Fees are not all due at once, but there is an initial outlay of about $350 to $500 for the average patent application which covers the preliminary search, drawings, attorney's fee, and government filing fee. More complicated patents, such as electronic devices, would have an initial cost closer to $750 to $1,000. Prosecution costs to be incurred at a later date would be extra as well as the government issuance fee.

Because a "lawyer's time is money," anything the inventor can do to make the job easier will be money saved. For instance, the drafting of the specification could probably be done essentially by the inventor and would save a few hours of the attorney's time. Some attorneys probably would agree to charge less because of this. Claims—the most important part of the patent document—should be written up by the attorney and rewritten by him during prosecution when challenged by the Patent Office.

Whether submitted by a patent attorney or the individual, certain forms must be included in the patent application. These consist of a written document comprising a petition, a specification (description and claims), and an oath or declaration that the applicant is the true and only inventor; drawings in those cases where required; and the government filing fee of $65 (plus an additional $10 for each independent claim in excess of the first).

The petition, specification, and oath or declaration must be in English and be legibly written in permanent black ink on one side of the paper (typewritten documents are preferred, of course). All parts should be sent in at once since an action cannot begin until every part is received. Drawings must be in black ink and rendered according to the rules prescribed by the Patent Office. Colored drawings, paintings, or color photographs are used only for design and plant patents.

How does one go about writing a patent application? The inventor first lists all the features and advantages of the invention, showing how it is an improvement over some other invention. (If a basic invention with no prior art, describe accurately just what it does.) Then list as many separate improvements as possible that the invention has over what is presently on the market. In a second list combine some of the separate improvements to make an even more improved invention. These single and combined improvements will constitute the claims of the patent and must later be drafted in legal language. (This will have to be done even if a patent attorney is handling the case for you, and the more complete the list is, the less time he will have to spend on the application, therefore allowing him to charge less.)

In order for the claims to be allowed by the Patent Office, subject matter of the claims must be supported by language having an antecedent basis in the specification or drawing.

The word "unobvious" is of prime importance in a chemical patent. If the reactions were old and well known and the results could be predicted in advance by anyone skilled in the art of chemistry, then there would be little chance of getting a patent. But, if something out of the ordinary resulted, then the chance for patentability would be greatly increased. The same would hold true for other types of inventions, i.e. an advance of knowledge that would be more than just "obvious" to someone skilled in the field.

The figures in a specification should show all the parts of the invention necessary to support the claim with each part numbered according to Patent Office drafting style and technique. A copy of the *Guide for Patent Draftsmen* (15¢) should be obtained if a person wishes to draw his own figures. Also in the Appendix is a list of Washington, D. C., firms that specialize in patent drawings if one wishes to have this work done for him.

When the invention is an improvement over an existing invention, great attention to detail of the improved area of the figure drawing must be made for this is the crux of the patent's approval or rejection.

The technique for writing claims is something that is acquired mainly by observation of issued patents, particularly claims of patents in one's own field of interest. Notice particularly how the words are put together, what key words sell the invention in legal language and make it patentable over the references cited against it. Most people have preferences for a particular field, be it electrical, chemical, mechanical, design or plant. This preference again narrows down to a still more specific field of interest. Consult the Patent Office Manual of Classifications or Classification Index for this particular class or subclass. Then browse through the Official Patent Gazette (found in most libraries) for the style used in the claims of your field of interest. In addition, you should order copies (50¢) of patents to see how the entire patent is composed.

The first paragraph of the patent application is now an abstract of the disclosure and the following material usually serves to orient the reader to the field in which the invention finds use. In it is explained the disadvantages of the present-day technology and the need for an invention to overcome these disadvantages.

The patent application then describes in general terms one of the features (objects) of the invention and how it overcomes these disadvantages (from your list of the invention's improvements). Subsequent paragraphs describe each of the other objects from your list. Next comes the general information part of the patent application: How to perform the objectives of the invention in the best possible way, other advantages, etc. Following this comes the examples or figures, or both, describing all the features of the invention so that anyone skilled in the art may likewise perform these same experiments and arrive at the same end results. Lastly, the claims protecting the objects of the invention are presented.

A complete patent application on a mechanical patent by one of Associated Ideas members (Mr. Harry Guedry) is presented to show the form of a cover letter addressed to the Commissioner of Patents; an Oath stating that Mr. Guedry believes he is the original inventor of a spring cutting device; a Petition that a patent be granted for his invention, and the actual patent application with appropriate marginal comments included for your information. (pp. 90-95.)

After these documents are forwarded they are given an official serial number by the Patent Office, which becomes the patent pending number and should appear on all further correspondence you have with the Office. The first action by the Office after acknowledgment of receipt of your patent application may take six months. But during this time, with a patent pending status, you have a measure of protection and may begin looking for a buyer to manufacture or even start selling the invention yourself. Remember, the full action on the patent probably will take at least two years, and it may be closer to three.

1809 North Broad Street
New Orleans, Louisiana 70119

April 15, 1965

Commissioner of Patents
U.S. Patent Office
Washington 25, D.C.

Dear Sir:

Enclosed you will find my patent application, "Improved Cutting Mechanism," an Oath and Petition, and a check for $30.00 (filing fee).

I wish to process this application myself; therefore, please address all correspondence to the above address.

Yours very truly,

Harry R. Guedry

HRG
ebr
encl.

(Note: Because of increased fees the cost to file this patent application with its two independent claims would now be $75. Under a new practice a declaration or oath may be filed.)

PETITION FOR PATENT

To the Commissioner of Patents:

Your petitioner, Harry R. Guedry, a citizen of the United States and a resident of New Orleans, State of Louisiana, whose post office address is 1809 N. Broad Street, prays that letters patent be granted to him for the improvement in a cutting mechanism, set forth in the following specifications.

OATH

I, Harry R. Guedry, being sworn, depose and say that I am a citizen of the United States and a resident of 1809 North Broad Street, New Orleans, Louisiana; that I verily believe myself to be the original, first and sole inventor of the invention "Cutting Mechanism" described and claimed herein; that I do not know and do not believe that the said invention was ever known or used before my invention thereof, or patented or described in any printed publication in any country before my invention therefore, or more than one year prior to said application, or in public use or on sale in the United States more than one year prior to said application; that said invention has not been patented before the date of said application in any country foreign to the United States on an application filed by me or my legal representatives or assigns more than twelve months prior to said application; and that no application for patent on said invention has been filed by me or my representatives or assigns in any country foreign to the United States.

WHEREFORE, I pray that Letters Patent be granted to me for the invention or discovery described and claimed in the foregoing specification and claims, and I hereby subscribe my name to the foregoing specification and claims, oath, and this petition.

_____

Harry R. Guedry

Sworn and subscribed before me
this   day of   , 1966

_____
Notary Public

IMPROVED CUTTING MECHANISM

Industry is approaching the vexing problem of cutting, clipping, shearing and macerating material of various composition by going to higher and higher horsepower and greater number of teeth on saws, clippers and blades. What with the need to frequently sharpen the teeth on these industrial devices the profit becomes less and less. Likewise, in other fields cutting becomes a compromise between quality, the need to sharpen less often and the danger of damaging items being out or clipped.

The object of this invention is to provide a cutter that does not need sharpening, uses less horsepower and cuts closer without damage; such an invention utilizes an extended spring or springs, the coils of which on closing cuts by crushing the object between said coils.

Fortunately, there is an endless number of types of springs made of various materials available in various tensions that would fulfill the object of this invention. To be operative the spring should be distended and allowed to pop back. Many ways would be applicable for carrying out the invention. One such way would be accomplished by attaching one end of the spring to an immovable section and the other end to a cam that on turning would quickly release the tension on the spring so that it returned with force. The cam would be attached to a prime mover for rapidity of cutting. Another way to accomplish the invention would be to use a vibrating mechanism; still another would be a solenoid to extend and release extension on the spring.

One or a plurality of springs would function to complete the objects of the invention. It would never be necessary to sharpen a cutting device having a cutting action from springs.

Still another object of this invention is that of selective cutting by limiting the space of the openings between each coil of the spring or springs.

Spring or springs would find application to the cutting of hair, threads, removal of cotton lint from cottonseed—to mention just a few uses. The invention would be accomplished wherever the cutting of an object capable of going into the opening between the coils of the spring—with the type of spring(s) and horsepower selected proportional to the strength of the material desired cut.

It would be recommended to use only a small part of the spring's total extension so as not to build up internal heat within the spring.

It is also recommended that the material be removed from the interior of the spring either by air or vibration or other satisfactory means.

The spring(s) would be placed in the open bottom of a hopper, or a basket of nothing but springs or the material could pass through the interior center of the spring(s). Auxiliary aids such as vibration, pressure, or air discharge on the material to be cut or shaved would function to complement the cutting action.

Any shaped spring with round, square, oblong, etc. coils would fulfill the object of the invention. Likewise the invention could be carried out with as little as one coil.

The following examples are illustrative of the invention and not to be considered as limitations thereof.

EXAMPLE 1

A spring was extended and hair allowed to fall within the coils. On popping back the hair was cut.

EXAMPLE 2

A spring was extended and threads were allowed to fall within the coils. On popping back the threads were cut.

EXAMPLE 3

Springs were placed side by side to enclose the bottom of an open chamber. The chamber was filled with cottonseed containing cotton linters thereon. The spring was extended so that there was a $\frac{1}{8}$ inch opening between each coil. The springs were then released simultaneously and it was noted that within the springs were cotton fiber linters indicative of selective cutting. These $\frac{1}{8}$ inch openings were sufficient to allow the fiber to go through but not the cottonseed itself.

EXAMPLE 4

The same conditions as in Example 3 were used except that the bank of springs lying side by side were immobilized and the free side attached to a straight edge which in turn was attached to a cam on the shaft of a motor. The mechanized assembly was run for 2 minutes and a much larger amount of cotton seed was delinted and a much larger amount of cotton linters was obtained than in Example 3.

The same device was used to shear paper, grind chemicals, strain trash from ground coffee, and blend and mix paint pigments.

Figure 4 illustrates a spring in an extended state with a limited space opening between each coil, and cottonseeds or like-type material seated on the spring with their associated linters falling through the openings between the coils of said spring.

Figure 5 illustrates the same spring as in Figure 2 in the next instant of time whereby the spring is in a nonextended state with the linters having been cut away by the crushing action of the spring on closing. This figure also illustrates the fact that there is no damage to the cottonseed itself because selective cutting has taken place by limiting the space opening between each coil of the spring.

Figure 6 is used only to illustrate a typical arrangement. Other cutting devices using the embodiment of the invention would be a hand shaver, a clothes or garment thread cutter, etc. This figure illustrates cutting, shearing, grinding, straining, blending and mixing applications whereby such items as cottonseed, paint piuments, chemicals and the like are placed in hopper 10, the bottom of which is composed of a plurality of springs 11, each lying one next to the other to completely enclose the opening in the bottom of the hopper, with one end of each spring 11 being attached to hopper 10 and the other end to rocker arm 12, said rocker arm attached to the main assembly at pivot point 13 and activated by cam 14 on shaft 15 with activation of the whole assembly by prime mover.

In other applications the hopper would be replaced by a belt or screw conveyor, if desired.

I claim the following:

1.   A method for cutting consisting of one or more springs the coils of which upon extension and nonextension cut the object or objects between said coils.

2.   A method for selective cutting by limiting the space of the openings between each coil of one or more extended springs, which upon nonextension cuts the object or objects between said coils.

3.   A method for eliminating sharpening of cutting surfaces in a cutting device by using one or more springs the coils of which upon extension and nonextension cuts the object or objects between said coils.

4. A method of shearing, grinding, straining, blending or mixing applications by using one or more springs the coils of which upon extension and nonextension shears, grind, strain, blend or mix the object or objects between said coils.

The passage of Public Law 89-83, approved on July 24, 1965, affects certain of the fees and matter relating to fees referred to in this publication. The new and revised fees payable under this Act are effective October 25, 1965. They are as follows:

FILING FEES

The filing fee of an application for an original patent, except in design cases, consists of a basic fee and additional fees. The basic fee is $65 and entitles applicant to present ten (10) claims, including not more than one (1) in independent form. An additional fee of $10 is required for each claim in independent form which is in excess of one (1) and an additional fee of $2 is required for each claim (whether independent or dependent) which is in excess of a total of ten (10) claims.

The following formula may be used in the calculation of the filing fee:

Basic Fee $65.00
Additional Fees:
Total number of claims in excess of 10,
  times $2................................. _____
Number of independent claims minus 1,
  times $10.............................. _____
    Total Filing Fee............... _____

patents will be treated in accordance with Public Law 89-83.

The issue fee for each original or reissue patent, except in design cases constitutes a basic fee of $100, and additional fees of $10 for each page (or portion thereof) of specification as printed, and $2 for each sheet of drawing.

The written notice of allowance given or mailed to each applicant entitled to a patent under the law will be accompanied by an estimate of the issue fee determined in accordance with the number of pages in the specification and the number of sheets of drawing. This issue fee is to be paid within three months thereafter and if timely payment is not made the application shall be regarded as abandoned (forfeited).

The notice of any remaining balance of the issue fee will be sent to the applicant at the time of the grant of the patent. If this remaining balance is not paid within three months therefrom the patent shall lapse.

DESIGN FEES

If an application for a design patent is received in the Patent Office after October 25, 1965, the fees prescribed by Public Law 89-83 are applicable.

On filing each design application, $20.

To avoid errors in the payment of fees it is suggested that a table such as given above be included in the letter of transmittal.

A claim is in dependent form if it incorporates by reference a single preceding claim which may be an independent or a dependent claim, and includes all the limitations of the claim incorporated by reference.

The Act also provides for the payment of additional fees on presentation of claims after the application is filed. This provision applies in the case of applications filed after October 25, 1965. When an amendment is filed which presents additional claims over the total number already paid for, or additional independent claims over the number of independent claims already accounted for, it must be accompanied by any additional fees due.

ISSUE FEES

Effective with notices of allowances mailed on and after October 25, 1965, the fees of issuing

On issuing each design patent: For three years and six months, $10; for seven years, $20; and for fourteen years, $30.

APPEAL FEES

A fee of $25 must accompany a notice of appeal filed for the first time when filed prior to October 25, 1965. If, in such a case, a brief is filed in support of the appeal after October 25, 1965, another fee of $50 is to be paid. If the notice of appeal is filed on or after October 25, 1965, the new fees prevail.

On appeal for the first time from the examiner to the Board of Appeals, $50; in addition, on filing a brief in support of the appeal, $50.

PATENT COPIES

Copies of plant patents in color, $1. Other uncertified copies of specifications and drawings of patents, 50 cents.

Figure 7. Notice of Increased Patent Office fees.

OTHER FEES

Without regard to the filing date of the application concerned, all disclaimers, petitions to revive or for the delayed payment of an issue fee received on or after October 25, 1965, must be accompanied by the fees provided for by Public Law 89-83.

On filing each disclaimer, $15.

On filing each petition for the revival of an abandoned application for a patent or for the delayed payment of the fee for issuing each patent, $15.

Similarly, requests for certificates under sections 255 or 256, or for recording of assignments, agreements, or other papers received on or after October 25, 1965, must be accompanied by fees in amounts provided for by Public Law 89-83.

For certificate under section 255 or under section 256 of this title, $15.

For recording every assignment, agreement, or other paper relating to the property in a patent or application, $20; where the document relates to more than one patent or application, $3 for each additional item.

[Applicable only when there is an identity in the assignor and assignee.]

TRADEMARK FEES

Section 3 of Public Law 89-83 increases certain fees payable to the Commissioner of Patents in connection with the registration of trademarks. The provisions of section 3 follow:

"(a) The following fees shall be paid to the Patent Office under this Act:

"1. On filing each original application for registration of a mark in each class, $35.

"2. On filing each application for renewal in each class, $25; and on filing each application for renewal in each class after expiration of the registration, an additional fee of $5.

"3. On filing an affidavit under section 8(a) or section 8(b) for each class, $10.

"4. On filing each petition for the revival of an abandoned application, $15.

"5. On filing opposition or application for cancellation for each class, $25.

"6. On appeal from the examiner in charge of the registration of marks to the Trademark Trial and Appeal Board for each class, $25.

"7. For issuance of a new certificate of registration following change of ownership of a mark or correction of a registrant's mistake, $15.

"8. For certificate of correction of registrant's mistake or amendment after registration, $15.

"9. For certifying in any case, $1.

"10. For filing each disclaimer after registration, $15.

"11. For printed copy of registered mark, 20 cents.

"12. For recording every assignment, agreement, or other paper relating to the property in a registration or application, $20; where the document relates to more than one application or registration, $3 for each additional item.

"13. On filing notices of claim of benefits of this Act for a mark to be published under section 12(c) hereof, $10."

Reverse side of Figure 7.

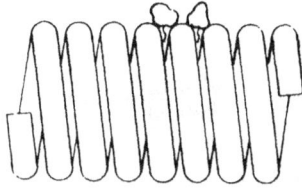

Figure 4.   Guedry cutting mechanism.

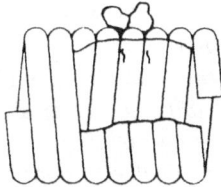

Figure 5.   As Figure 4, with spring nonextended.

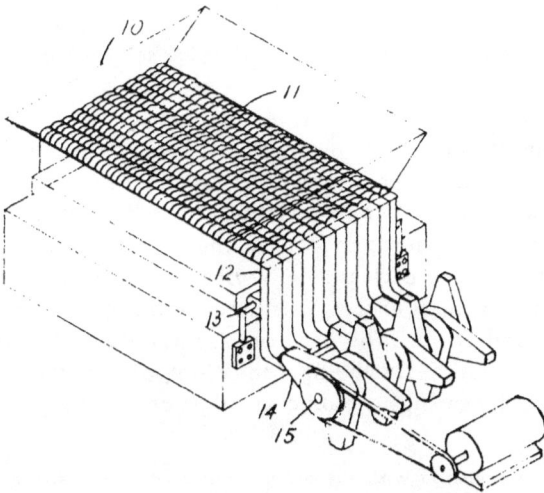

Figure 6.   Cutting mechanism in practical application.

VI. Patent Office Prosecution

It is not unusual for the inventor to have to wait nine months for the first official action from the Patent Office. When it does come up for investigation, a thorough analysis is given to it. The search by the Patent Office is probably much more thorough than that done by the inventor or his representative. Not only are all prior patents checked, but also reams of recently published material that might have bearing on the case. One inspector even quoted an article in Ripley's "Believe It or Not" as evidence that the device in question had appeared in print more than one year before patent application date.

Patents are granted for about four out of every seven applications filed, according to the Patent Office. This is a high rate considering that many patents are submitted by persons who have not really determined the "uniqueness" of their idea, but cannot be dissuaded from filing application. With this percentage of "patents that shouldn't have been submitted in the first place" taken into account, the chances of a worthwhile patent being accepted are closer to five out of seven— good odds.

But the possibility of a patent going through without being challenged is almost nil. And, more than likely, claims will be turned down two or three times before they satisfy the examiner's criteria.

In many cases the patent is declared rejected on informal grounds. This is the result of improper form in the application papers or drawings. Often, it is the oath that is wrong—it is not copied correctly, there are typographical errors, it is dated and notarized more than 90 days before application. The petition also has a stated form that must be followed minutely, including the all-important clause about the invention not appearing in public print prior to one year before ap-

This chapter was prepared with the aid of registered patent attorney Mr. Calvin J. Laiche, Trombatore, Vonderstein and Laiche, Kenner, Louisiana.

plication. In the cases of informal grounds for rejection the Patent Office will spell out just what is wrong and what must be done to correct it. The patent will still be processed, however, while these informal amendments are being corrected. Most of the time the notation of informal amendments will be held and sent back to the inventor along with the first official action of formal grounds for rejection of claims.

The specifications will not be challenged unless something appears in the claims that is not "anticipated" in the specifications. Usually they remain untouched and become a matter of public record upon issuance. This can be important because a claim might be disallowed as far as inventor's exclusive rights are concerned, but the same example appearing in the specifications will thus mean that no one else can get a patent on this particular use because it becomes public knowledge. The inventor may lose a certain portion of his rights, but he also keeps anyone else out of the field as far as their getting exclusive rights on that portion which he lost. Thus in Guedry's patent, one of the claims for using the cutting mechanism as a mixer was disallowed, but as it appears in the specifications, a mixer manufacturer cannot use it as well, except as common knowledge that can be duplicated by anyone else in the field. This is no real advantage to the inventor, but the primary reason for patents is for the public benefit and the advancement of knowledge.

On the first formal grounds of rejection of claims, the Patent Office will probably cite four to eight patents to prove that your application shouldn't be patented. You will usually have six months to answer this first rejection, but subsequent action may be shortened to a period of 90 days. In that time you will have to study copies of the patents cited against you and show just how your patent is sufficiently different to warrant issuance. A mere denial of the examiner's position is not enough. Claims will have to be rewritten, some may be entirely lost because they cover too much of the field, but that is par for the course as patents are usually submitted to get the broadest claims possible, with the knowledge that they face the likelihood of being trimmed back.

There are three general areas of rejection under the United States Code of Patent Law that could be cited against your patent.

If the Patent Office makes reference to 35 U.S.C. 102 in its rejection it is classifying your invention as not being "new" enough. Most things are new in some respect, but if your invention is basically the same as another except in minor details—i.e. its trimming or "hardware"—

you will have little chance to win your case. In cases like this you cannot "argue around" another patent's disclosure because it meets yours head-on—your classification falls right in the middle of another's disclosure field. The only recourse left would be what is called a "picture" patent—a weak patent that is almost forced through on the basis of extra trimmings that are not really necessary and relevant to the patent's intent.

35 U.S.C. 101 covers those inventions deemed frivolous—not useful in any way. Considering the wide variety of gimmicks, toys, luxury party-type products on the market, the Patent Office could be said to be very lenient in this respect. Almost anything of marketable value can be shown useful for some person. It is a grounds for rejection seldom brought up except in some chemical or medicinal art areas or perpetual motion "inventions."

A much more difficult rejection reference (both for the Patent Office and the inventor) is 35 U.S.C. 103 wherein the "nonobviousness" of the invention is questioned. This means that in light of previous knowledge, by one versed in the field, your invention would have been anticipated. Thus an expert would know beforehand how your process worked, or what your mechanical improvement was, but that it was not sufficiently different from known processes to bother patenting it —it was just an extension of knowledge already in use. In this case hindsight is always easier than foresight. Once another person sees and understands your idea for an invention, he might say "well, of course, that is the way it would have to be done." But, in reality, he didn't think about it, the inventor was the one who created and perfected a new idea—an idea that was easy to explain only after it had been worked out in detail. If your idea was substantially "taught" in the field, then the objection is very strong. But if only "anticipated," you have grounds to argue on, and can proceed to prove how you materially advanced the "state of the arts" in a way that had not been thought of before.

When your patent is rejected under 35 U.S.C. 103 with phrases such as "unpatentable in light of," "taken in connection with," "lacking invention over" such and such prior art, then you know that the Patent Office is hedging somewhat. They feel your case has some merit and can not pin down a specific reference to cite against you and so quote a half a dozen patents that are closely related to yours. Then it is up to you to rewrite your claims so they do not overlap those of the cited patents, and yet still keep as strong a protection as possible for your own invention. A second rejection will probably follow from the Patent Office notifying you to further particularize your claims in some respect, and it will be labeled a final action; at this stage you will either receive your patent or be forced to take the application to an appeal board.

Patent Office action on the "cutting mechanism" patent application of Harry Guedry was fairly typical for mechanical inventions. Their first reply, after acknowledgment of receipt, was eight months in coming. In it technical changes were asked in the oath (and these were spelled out in detail for the benefit of the inventor) and all four claims were rejected. Reply time was set at three months and all subsequent letters made the same notation as the Patent Office attempted to speed up its operation.

Two patents, Lanzisera 2,496,223 and Forward 3,194,001, were used as grounds for rejection. The statement read: "Claims 1-4 are rejected under 35 USC 102 as being obviously fully met by Forward. The reference to Lanzisera is cited to show a similar use for spring members." The patent held by Forward described a thread cutter that was a non-reciprocating device wherein a thread running through a spring was cut at a certain tension and place by an arm arrangement that had to be reset for each new cutting.

Lanzisera was a chicken plucker! The springs in this case surrounded the chicken and grabbed the feathers while another action pulled the chicken away from the springs to actually cause the plucking action.

Answering these rejections was fairly easy, but it meant losing the broad coverage of the mechanism for use in shearing paper, grinding chemicals, straining trash from coffee, and blending and mixing paint pigments. A new claim was submitted in place of the four that were rejected, and this time only the use of the mechanism for cutting fibrous materials (such as cotton or hair) was kept.

Guedry's answer to the Patent Office rejection read:

> Neither reference contemplates the device of newly presented claim 5. U.S. 2,496,223 is not a cutting mechanism and the device of U.S. 3,194,001 would require radical modification in order to operate according to the device of instant application. The Forward reference is entirely lacking in the concept of reciprocating mechanism.

A new claim No. 5 was submitted and worded in the technical jargon of the trade to read:

> 5. Apparatus for cutting particulate, fibriform materials, the individual elements of which are substantially equable in cross section comprising: at

least one coil spring defined by a plurality of integral coil members, each coil member, by virtue of the spring design, normally pressed tightly against the adjacent counterpart coil member; reciprocating means affixed to the ends of said coil spring and adapted (a) to extend said spring and separate adjacent coil members a distance at least sufficient to pass individual elements of the particular fibriform material through the interstices thus formed, and (b) to release said coil spring when extended, the speed of release being at least equal to the inherent recovery speed of the said spring.

All of which means simply that the spring opens up to let some kind of fiber or thread pass through and then snaps close, with the edges of the spring acting like knives to cut the fiber. It is done in a continuous open and closing (reciprocating) action driven by some means to extend and release the spring or body of springs.

The second office action came three months later and was a rejection of claim 5 and this was final! But they left a loophole pointing the way to a satisfactory amendment.

The pertinent paragraphs read:

Claim 5 is rejected under 35 U.S.C. 103 and unpatentable over Forward. Therein is a taut coil spring 11 in which the coils are normally pressed tightly together and a reciprocating means (linkage 15, 16, 18) which actuates the slides 13 along rod 14 to effectuate flexing of the coils.

"The above rejection could be overcome by a more specific recitation of the structure as shown in Figure 3 of the drawings.

Remember, Forward was the thread cutting device that most nearly approached Guedry's invention but it was limited to just one thread and the action was not really continuously reciprocating in that it had to be reset each time a new thread was introduced.

Guedry's answer is given in full to point out two important points in filing amendments. First, make reference to the specific rejection and answer it, and give the corrections that must be made in the claims. Then rewrite the *whole* corrected claim. Because claims often have to be rewritten and reference made to a specific insertion point, it might be easier to have numbered lines on the margins of your original claim sheet so you can refer to the line and the paragraph.

In re Application of
Harry R. Guedry
Serial No. 468,193
Filed June 28, 1965

The Honorable
The Commissioner of Patents

Sir:

This is in response to the Examiner's final action dated May 31, 1966.

To the end that the final rejection of Claim 5, the only remaining claim in the case, might be overcome by "a more specific recitation of the structure," as kindly suggested by the Examiner, please amend the subject case as follows:

IN THE CLAIMS

In Claim 5, after "coil member," insert—support means fixed to each end of said coil spring and adapted to locate and support said coil spring. In Claim 5, cancel "affixed to the ends of said coil spring and," and insert in place thereof—with associated drive means therefore located adjacent at least one of said support means, said reciprocating means.

REMARKS

Claim 5 amended as above will then read as follows:

5. Apparatus for cutting particulate, fibriform materials, the individual elements of which are substantially equable in cross section comprising: at least one coil spring defined by a plurality of integral coil members, each coil member, by virtue of the spring design, normally pressed tightly against the adjacent counterpart coil member; support means attached to each end of said coil spring and adapted to locate and support said coil spring; reciprocating means with associated drive means therefore located adjacent at least one of said support means, said reciprocating means adapted (a) to extend said spring and separate adjacent coil members a distance at least sufficient to pass individual elements of the particulate fibriform material through the interstices thus formed, and (b) to release said coil spring when extended, the speed of release being at least equal to the inherent recovery speed of the said spring.

The answer was simply a more restrictive description of the means of reciprocating action so as not to conflict with the particular mechanical drive action of the thread cutting mechanism. The new amendment defined more clearly just what type of machine action would have to be used in producing the cutting mechanism, but it did not change the basic idea.

In addition, certain other minor changes were made by the Patent Office with the stipulation that if they were not acceptable, another amendment would have to be filed. The changes were the labeling on the drawing page of what the figures actually represented, i.e. "Figure 4 is a view of one of the springs in the extended state." In line 21 of page 3 of the specification the phrase "Figure 4 illustrates a spring" has been cancelled and the following phrase inserted before the word "in"—Referring to Figure 4 a spring is illustrated. As these changes were so minor, no new amendments were filed and the Patent was issued 16 months after the first submission.

If the Patent Office examiner turns down all amendment applications and insists that the invention will not be patented, the inventor still has recourse to appeal. His first appeal is to the Patent Office Board of Appeals. If unsuccessful there, he may sue the Commissioner of Patents in the District Court of Washington, D. C., or he may appeal to the Court of Customs and Patent Appeals. Both actions have an even longer waiting period and are extremely costly in attorney's fees, printing up of the record of the case, and expert witnesses that may be needed in your behalf. The services of an attorney of the Court of Customs and Patent Appeals would have to be used in the District Court, but a Patent Agent or the inventor can file an appeal to the Patent Office Board of Appeals. Besides the cost and time involved, both choices have certain inherent disadvantages that must be weighed against the proposed value of the patent if it is issued.

In suing in the District Court, the regularly appointed judges preside over the case. Because they are not intimately aware of all the technical points involved in patent applications, they may favor the plaintiff if he can "gloss" over the weak points in his case. The case will probably come up quicker, but all records are open to the public, and if there is anything that needs to be kept secret, it may be lost. In addition, the Patent Office can challenge the patent application on any grounds, not just final rejection. Thus, new evidence may be introduced that was not even considered before and the inventor may have to defend his position against a whole new line of attack.

The Patent Office Board of Appeals is made up of the Commissioner of Patents, The Assistant Commissioners, and usually three other examiners-in-chief.

A brief will have to be filed in this case and the inventor may appear in person if he desires, but it is not absolutely necessary. The examiners—acknowledged experts in the field—will only consider the action on the final rejection, and judge whether the grounds of rejection—be it prior art, other patents, etc.—were sufficient to turn down the application. The inventor in turn can not bring in any new evidence except, of course, his arguments for the merits of the case. Whatever the decision rendered, it is final.

In rare cases, the Patent Office may determine that two applications are so similar that interference proceedings must be instituted. An inventor can also claim interference proceedings within one year of his patent issuance if, for instance, he sees in the Official Gazette a description of an invention that he believes is essentially like his. In this case, he submits a patent application with exactly the same claims as the one in the Official Gazette, and states that he wants the Patent Office to start interference proceedings. The Patent Office bears the cost of this, except attorney fees for the inventor.

Interference means that the inventor must prove that he was the first to conceive the invention and reduce it to practice by actual construction, operation, testing, or disclosure. Every date the inventor has kept to show his continued work on the invention is extremely important, and any dates left out of the records may cost the inventor his patent. If the inventor sees a "carbon copy" of his work in the Official Gazette, he can have an affidavit notarized that states he thought of the invention before the date of filing on the other patent. He does not reveal his own date of conception because that might give advantage to the other party. He still must prove by his records that he had conceived of the idea before the filing date of the other patent. Many interference cases are settled by the inventors through cross-license agreements for royalty rights so that both parties can get some benefit from their work. Now, the law requires notification to the Patent Office of just what settlement terms were made.

While interference is an intra-office matter within the Patent Office, infringement means the suing of another the Federal courts for the unlawful use, making, or selling of a patented invention. An injunction may be sought to make the other party cease manufacture of the invention in question, and financial compensation can be gained. Many infringement cases involve determination of who has the dominant patent as outlined in the claims. As stated before, an invention may be patentable but still fall under domination of a stronger patent that is unexpired. In case of infringement, the other party should be put on notice that you may sue. Often a license or royalty agreement is worked out depending upon who has the best patent. The court costs * in this type of case can be very high; thousands of dollars could be spent, and the case still lost. No matter what, a Patent Attorney would have to represent the inventor, and his decision would be based on the relative strength or weakness of the patent claims.

Patent rights extend only throughout the United States and if an inventor wants protection in another country, he will have to file for application in that particular country within one year of applying for his own patent. The laws of each country are different and some require that an invention actually be manufactured there after a certain period of time (usually three years), or a penalty tax will be placed on the patent. The Patent Office will not assist in the filing of an application in a foreign country, but your patent attorney or agent can directly or indirectly do this for you. Some patent brokers will also take on the burden of the fees for a percentage of profits. While the customs, habits, and manufacturing abilities of foreign countries must be taken into account, there are some inventions that are profitable almost anywhere in the civilized world. One example of this is the foreign patent filed by American inventors and used by the Proban Company in England for flame-proofing material. Recently England passed a law declaring that all material, such as curtains, clothing, bed spreads, etc. must be flame-proofed. Proban, which has one of the best processes for this, is almost assured of royalty rights by law—an enviable position, to say the least.

ANSWERS TO QUESTIONS FREQUENTLY ASKED ABOUT PATENTS

Q. Will a letter to yourself protect you?

A. Many persons believe that they can protect their inventions against later inventors merely by mailing themselves a registered letter describing the invention. This is not true. Your priority

*It has been estimated that the average patent suit costs each party $50,000.

rights against anyone else who makes the same invention independently cannot be sustained except by testimony of someone else who corroborates your own testimony as to all the important facts such as conception of the invention, diligence, and the success of any tests you may have made. It is therefore important that some trustworthy friend witness these things. The invention will not be fully protected until patented.

Q. Will I automatically receive copies of all correspondence between my attorney and the Patent Office?

A. You should request your attorney to send you copies of all correspondence in order to be informed and possibly to help your attorney in his effort to overcome references cited against your patent application.

Patent Applications

Q. I have made some changes and improvements in my invention after my patent application was filed in the Patent Office. May I amend my patent application by adding a description or illustration of these features?

A. No. The law specifically provides that new matter shall not be introduced into the disclosure of a patent application. However, you should call the attention of your attorney or agent promptly to any such changes you may make or plan to make, so that he may take or recommend any steps that may be necessary for your protection.

Q. How does one apply for a patent?

A. By making the proper application to the Commissioner of Patents, Washington, D. C. 20231.

Q. Of what does a patent application consist?

A. An application fee, a petition, a specification and claims describing and defining the invention, an oath or declaration, and a drawing if the invention can be illustrated.

Q. Are models required as part of the application?

A. Only in the most exceptional cases. The Patent Office has the power to require that a model be furnished, but rarely exercises it.

Q. Is it necessary to go to the Patent Office in Washington to transact business concerning patent matters?

A. No. Most business with the Patent Office is conducted by correspondence. Interviews regarding pending applications can be arranged with examiners if necessary, however, and are often helpful.

Q. Can the Patent Office give advice as to whether an inventor should apply for a patent?

A. No. It can only consider the patentability of an invention when this question comes regularly before it in the form of a patent application.

Q. Is there any danger that the Patent Office will give others information contained in my application while it is pending?

A. No. All patent applications are maintained in the strictest secrecy until the patent is issued. After the patent is issued, however, the Patent Office file containing the application and all correspondence leading up to issuance of the patent is made available in the Patent Office Search Room for inspection by anyone, and copies of these files may be purchased from the Patent Office.

Q. May I write to the Patent Office directly about my application after it is filed?

A. The Patent Office will answer an applicant's inquiries as to the status of the application, and inform him whether his application has been rejected, allowed, or is awaiting action by the Patent Office. However, if you have a patent attorney or agent the Patent Office cannot correspond with both you and the attorney concerning the merits of your application. All comments concerning your invention should be forwarded through your patent attorney or agent.

Q. Can the 6-month period allowed by the Patent Office for response to an office action in a pending application be extended?

A. No. This time is fixed by law and cannot be extended by the Patent Office. The application will become abandoned unless proper response is received in the Patent Office within this time limit.

When to Apply For Patent

Q. I have been making and selling my invention for the past 13 months and have not filed any patent application. Is it too late for me to apply for patent?

A. Yes. A valid patent may not be obtained if the invention was in public use or on sale in this country for more than one year prior to the filing of your patent application. Your own use and sale of the invention for more than a year before your application is filed will bar your right to a patent just as effectively as though this use and sale had been done by someone else.

Q. I published an article describing my invention in a magazine 13 months ago. Is it too late to apply for patent?

A. Yes. The fact that you are the author of the article will not save your patent application. The law provides that the inventor is not entitled to a patent if the invention has been described in a printed publication anywhere in the world more than a year before his patent application is filed.

Who May Obtain a Patent

Q. Is there any restriction as to persons who may obtain a United States patent?

A. No. Any inventor may obtain a patent regardless of age or sex, by complying with the provisions of the law. A foreign citizen may obtain a patent under exactly the same conditions as a United States citizen.

Q. If two or more persons work together to make an invention, to whom will the patent be granted?

A. If each had a share in the ideas forming the invention, they are joint inventors and a patent will be issued to them jointly on the basis of a proper patent application filed by them jointly. If on the other hand one of these persons has provided all of the ideas of the invention, and the other has only followed instructions in making it, the person who contributed the ideas is the sole inventor and the patent application and patent should be in his name alone.

Q. If one person furnishes all of the ideas to make an invention and another employs him or furnishes the money for building and testing the invention, should the patent application be filed by them jointly?

A. No. The application must be signed, sworn to, and filed in the Patent Office in the name of the true inventor. This is the person who furnishes the ideas, not the employer or the person who furnishes the money.

Q. May a patent be granted if an inventor dies before filing his application?

A. Yes. The application may be filed by the inventor's executor of administrator.

Q. While out of the country I find an article on sale that is very ingenious and has not been introduced into the United States or patented or described. May I obtain a United States patent on this invention?

A. No. A United States patent may be obtained only by the true inventor, not by someone who learns of an invention of another.

Ownership and Sale of Patent Rights

Q. May the inventor sell or otherwise transfer his right to his patent or patent application to someone else?

A. Yes. He may sell all or any part of his interest in the patent application or patent to anyone by a properly worded assignment. The application must be filed in the Patent Office as the invention of the true inventor, however, and not as the invention of the person who has purchased the invention from him.

Q. If two persons own a patent jointly, what can they do to grant a license to some third person or company to make, use or sell the invention?

A. They may grant the license jointly, or either one of them may grant such a license without obtaining the consent of the other. A joint owner does not need to get the consent of his co-owner either to make, use, or sell the invention of the patent independently, or to grant licenses to others. This is true even though the joint owner who grants the license owns only a very small part of the patent. Unless you want to grant this power to a person to whom you assign a part interest, you should ask your lawyer to include special language in the assignment to prevent this result.

Q. As joint inventor, I wish to protect myself against the possibility that my co-inventor may, without my approval, license some third party under our joint patent. How can I accomplish this?

A. Consult your lawyer and ask him to prepare an agreement for execution by you and your co-inventor to protect each of you against this possibility.

Duration of Patents

Q. For how long is a patent granted?

A. Seventeen years from the date on which it is issued, except for patents on ornamental designs, which are granted for terms of $3\frac{1}{2}$, 7, or 14 years.

Q. May the term of a patent be extended?

A. Only by special act of Congress, and this occurs very rarely and only in most exceptional circumstances.

Q. Does the patentee continue to have any control over use of the invention after his patent expires?

A. No. Anyone has the free right to use an invention covered in an expired patent so long as he does not use features covered in other unexpired patents in doing so.

Patent Searching

Q. Where can a search be conducted?

A. In the Search Room of the Patent Office in Capital City, Virginia. Classified and numerically arranged sets of United States and foreign patents are kept there for public use.

Q. Will the Patent Office make searches for individuals to help them decide whether to file patent applications?

A. No. But it will assist inventors who come to Washington by helping them to find the proper patent classes in which to make their searches. In response to mail inquiries it will also advise inventors as to what patent classes and subclassds to search. For a reasonable fee it will furnish lists of patents in any class and subclass, and copies of these patents may be purchased for 50 cents each.

Attorneys and Agents

Q. Does the Patent Office control the fees charged by patent attorneys and agents for their services?

A. No. This is a matter between you and your patent attorney or agent in which the Patent Office takes no part. In order to avoid possible misunderstanding you may wish to ask him for estimates in advance as to his approximate charges for (a) the search, (b) preparation of the patent application, and (c) Patent Office prosecution.

Q. Will the Patent Office inform me whether the patent attorney or agent I have selected is reliable or trustworthy?

A. All patent attorneys and agents registered to practice before the Patent Office are expected to be reliable and trustworthy. The Patent Office can report only that a particular individual is or is not in good standing on the register.

Q. If I am dissatisfied with my patent attorney or agent, may I change to another?

A. Yes. There are forms for appointing attorneys and revoking their powers of attorney in the pamphlet entitled "General Information Concerning Patents."

Q. How can I be sure that my patent attorney or agent will not reveal to others the secrets of my invention?

A. A patent attorney and agent earns his livelihood by the confidential services he performs for his clients and if any attorney or agent improperly reveals an invention disclosed to him by a client, that attorney or agent is subject to disbarment from further practice before the Patent Office and loss of his livelihood. Persons who withhold information about their inventions from their attorneys and agents make a serious mistake, for the attorney or agent cannot do a fully effective job unless he is fully informed of every important detail.

Plant and Design Patents

Q. Does the law provide patent protection for invention of new and ornamental designs for articles of manufacture, or for new varieties of plants?

A. Yes. If you have made an invention in one of these fields, you should read the Patent Office pamphlet "General Information Concerning Patents."

Technical Knowledge Available From Patents

Q. I have not made an invention but have encountered a problem. Can I obtain knowledge through patents of what has been done by others to solve the problem?

A. The patents in the Patent Office Search Room in Washington contain a vast wealth of technical information and suggestions, organized in a manner which will enable you to review those most closely related to your field of interest. You may come to Washington and review these patents, or engage a patent practitioner to do this for you and to send you copies of the patents most closely related to your problem.

Infringement of Other's Patents

Q. If I obtain a patent on my invention will that protect me against the claims of others who assert that I am infringing their patents when I make, use, or sell my own invention?

A. No. There may be a patent of a more basic nature on which your invention is an improvement. If your invention is a detailed refinement or feature of such a basically protected invention, you may not use it without the consent of the patentee, just as no one will have the right to use your patented improvement without your consent. You should seek competent legal advice before starting to make or sell or use your invention commercially, even though it is protected by a patent granted to you.

Enforcement of Patent Rights

Q. Will the Patent Office help me to prosecute others if they infringe the rights granted to me by my patent?

A. No. The Patent Office has no jurisdiction over questions relating to the infringement of patent rights. If your patent is infringed you may sue the infringer in the appropriate United States court at your own expense.

Patent Protection in Foreign Countries

Q. Does a United States patent give protection in foreign countries?

A. No. The United States patent protects your invention only in this country. If you wish to protect your invention in foreign countries, you must file an application in the Patent Office of each such country within the time permitted by law. This may be quite expensive, both because of the cost of filing and prosecuting the individual patent applications, and because of the fact that most foreign countries require payment of taxes to maintain the patents in force. You should inquire of your practitioner about these costs before you decide to file in foreign countries.

National Defense Inventions

Q. I believe that the publication of a granted patent on my invention would be detrimental to the national defense. For this reason, I am reluctant to file a patent application unless there is a special method of handling cases of this nature. What should I do.

A. You need have no qualms about filing an application in the U. S. Patent Office. If it is determined that publication of the invention by granting of a patent would be detrimental to the national defense, the Commissioner of Patents will order that the invention be kept secret and will withhold the grant of a patent until such time as a decision is made that disclosure of the invention is no longer deemed detrimental to the national security. If an order is issued that the invention of your patent application be kept secret, you will be entitled to apply for compensation for any use of the invention by the Government, if and when the application is held to be allowable.

Developing and Marketing Inventions and Patents

Q. Will the Patent Office advise me as to whether a certain patent promotion organization is reliable and trustworthy?

A. No. The Patent Office has no control over such organizations and cannot supply information about them. It is suggested that you obtain this information by inquiring of the Better Business Bureau of the city in which the organization is located, or of the Bureau of Commerce and Industry of the state in which the organization has its place of business. You may also undertake to make sure that you are dealing with reliable people by asking your own patent attorney or agent whether he has knowledge of them, or by inquiry of others who may know them.

Q. Are there any state government agencies that can help me in developing and marketing of my invention?

A. Yes. In nearly all states there are state planning and development agencies or departments of commerce and industry that are seeking new product and new process ideas to assist manufacturers and communities in the state. If you do not know the names or addresses of your state organizations you can obtain this information by writing to the Governor of your state.

Q. Can the Patent Office assist me in the developing and marketing of my patent?

A. Only to a very limited extent. The Patent Office cannot act or advise concerning the business transactions or arrangements that are involved in the development and marketing of an invention. However, the Patent Office will publish, at the request of a patent owner, a notice in the "Official Gazette" that the patent is available for licensing or sale. The fee for this service is $3.

Q. Can any U. S. Government agency other than the Patent Office assist me in the development and marketing of my invention?

A. The Business and Defense Services Administration of the U. S. Department of Commerce, Washington 25, D. C., may be able to help you with information and advice, as its various industry divisions maintain close contact with all branches of American industry, or you may get in touch with one of the Department of Commerce field offices.

Proposed Patent Law Changes *

The patent laws are enacted by Congress in accordance with the power granted it by the Constitution (Article I, Section 8, clause 8), which provides:

> The Congress shall have the Power......to promote the Progress of Science and useful Arts by Securing for Limited Times to Authors and Inventors the Exclusive Right to their Respective Writings and Discoveries.

*By C. Emmett Pugh, Esq., former Patent Examiner and registered patent attorney, partner in the firm of Drury, Lake & Pugh. The writer would like to acknowledge the generous assistance in this matter of Edward F. McKie, Jr., Esq., Chairman of the Patent, Trademark and Copyright Law Section of the American Bar Association. See end of chapter for notes.

In 1952[1] Congress for the first time since 1870[2] completely rewrote the patent statutes. Approximately six years later, Congress began again to study and consider sweeping changes in the provisions of the federal patent laws.

On April 8, 1965, during the celebration of the 175th[3] anniversary of the patent system, President Lyndon B. Johnson issued Executive Order No. 11,215 establishing the President's Commission on the Patent System to study and make recommendations for improving the patent system. The Commission in its report[4] made some 35 recommendations, a number of which suggested vast and fundamental changes.

Among the major changes proposed was the elimination of the present interference proceeding[5] and in its place the substitution of a first-to-file system. Thus, an applicant who filed his patent application first would be granted the patent although another, who was claiming the same invention but who filed later, was the first inventor (first to conceive and/or first to actually reduce the invention to practice). In effect, this creates a "race to the Patent Office."

Another major proposal of the President's Commission, designed to speed the patent disclosure to the public-the quid pro quo for the patent grant[6]—was the publication of each application no later than 18-24 months from its filing. Since the proposed forced publication might occur before the applicant had a full determination of the patent scope to be awarded (if any),[7] this proposal has met some opposition. Heretofore an applicant, if he felt he was *not* going to be awarded a sufficient scope of patent protection, had the option of abandoning his application and maintaining his invention as a trade secret. Under the proposal of the President's Commission the inventor thus may give his part of the bargain (public disclosure) without necessarily receiving the corresponding patent grant.

Other major proposals included the elimination of the present one-year grace period[8]; the addition of use, knowledge, and sale in *foreign countries* to the art barring the issuance of a patent[9]; changing the patent grant term from seventeen to twenty years; the elimination of design and plant patents[10]; and allowing the filing of a patent application by the owner thereof, vis-a-vis the inventor himself.[11]

The various proposals of the President's Commission, within three months of the issuance of the report, were drafted into legislation and introduced in both Houses of Congress by the Johnson Administration. Hearings on the proposed legislation began in the House of Representatives and the Senate on April 17 and May 17, 1966, respectively.

From the time of the Commission report and the introduction of its legislative embodiment, much study and assistance have been generated by such knowledgeable and concerned bodies as the American Patent Law Association, the Patent, Trademark and Copyright Law Section of the American Bar Association, the National Association of Patent Law Associations, the National Association of Manufacturers (NAM), the United States Chamber of Commerce, the American Chemical Society and, of course, the United States Patent Office itself.

As a result of the legislative hearings, the public discussions that ensued, and the introduction of other legislative bills,[12] the Administration has dropped or substantially altered many of the Commission's proposals.[13] In particular it now appears that there will not be an adoption of a first-to- file system but rather a modification of the present interference[14] provisions. Also abandoned has been the attempt to eliminate the one-year grace period.

Although nowhere near as revolutionary as the initial proposal, apparently some substantial changes are going to be made by the Congress to better the patent system. The following indicates what most likely the changes will be.

The priority (nee interference) proceedings will be limited in some way in time to one or two years, either as a maximum permissible time difference in filing dates or as a limitation as to how far a party can go back with its proofs. In addition, both parties, in establishing a date of invention earlier than their filing dates, must show due diligence on their part up to the filing of the applications—unwarranted delays will be fatal. Moreover, rather than have a priority contest invoked between two pending applications, the Patent Office will probably issue the earlier filed case and, upon issuance, use it as an anticipating reference against the later filed case. The latter then has the tri-part option of invoking an interference with the patent by copying its claims, restricting its claims to patentably distinguish over the patent (assuming he has such further limitations disclosed) or, of course, abandoning the application.[15] All of the proposed changes, particularly the time limitation, will to some degree simplify the priority-contest procedure and hence be of some assistance to the independent inventor. To succeed under the present system one was usually required to endure and participate in protracted and complex (and correspondingly very expensive) proceedings.

Congress will likely provide for the publication of applications prior to their issuance as patents but such publication will be voluntary, not forced, and will be allowed any time at the election of the applicant. As an incentive to publication, the law will allow recovery for infringement damages from the first date after both publication and actual notice of the applicant's claim to the infringer.[16] A further advantage of publication is to bar[17] anyone else from patenting the same invention more than a year later, even though the publishing applicant abandoned his application.

Also under strong consideration is the creation of a patent revocation proceeding wherein an issued patent may later be declared invalid or caused to be further restricted by the Patent Office.

The procedure being studied would permit the citation of additional prior art by members of the public up to a year after the issue date, upon which the Patent Office in its discretion could then require the patentee to show cause why the patent should not be declared invalid or, alternatively, further restricted in scope. Such a procedure would be of tremendous benefit to the independent inventor as it would allow him to in effect invalidate or restrict a contrary patent without incurring the tremendous cost[18] of personally fighting a patent law suit[19] in the federal courts.

It appears to be almost definite that the term of the patent grant will be changed from seventeen to twenty years, but that the term will then be measured from the *filing* rather than the *issue* date. The latter change will encourage the quickest of action on the applicant's part as he prosecutes his application through the Patent Office. Heretofore, if there were no infringer on the scene, an applicant would tend to drag out the prosecution to thereby defer as much as possible the expiration date of his envisioned patent.

All the changes in the offing will, it is believed, serve to better this country's patent system and be of assistance to the independent inventor—a participant in the patent system too often overlooked.

Notes

1. Patent Act of July 19, 1952 (Public Law 593, 82nd Cong., 2nd sess., ch. 950, 66 Stat. 792).
2. Act July 8, 1870, c. 230, 16 Stat. 198.
3. The first patent law was approved in 190 (Act April 10, 1790, c. 7, 1 Stat. 109).
4. Report released December 8, 1966; a copy of the report may be obtained from the Government Printing Office (90th Cong., 1st sess., Senate Docu. No. 5; 74-184).

5. A proceeding before the Patent Office for determining priority between two or more inventors claiming the same invention; this proceeding grants the patent to the applicant who was the first to make the invention.

6. In essence a patent is a contract between the government, acting on behalf of the public, and the patentee in which the patentee is granted a seventeen-year exclusive monopoly in return for a public disclosure of the invention.

7. Because of a backlog of over 200,000 applications in the Patent Office, an applicant usually waits anywhere from six to eighteen months before receiving a first Office Action or substantive response to his application; a full determination of the patent scope does not normally come until at least the second Office Action, which usually comes six months to a year later.

8. Section 102 of the patent statutes (Title 35, United States Code) permits an inventor up to *one year* to file for patent protection after the invention has been (a) patented (for example, in a foreign country), or described in a printed publication anywhere, (b) publicly used and known in this country, or (c) put on sale in this country.

9. Note sections (b) and (c) of note 8 above.

10. The patent statutes heretofore had allowed patents to be granted for new, oriinal and ornamented designs of articles for optional periods of three-and-a-half, seven or fourteen years, and for new and distinct asexually reproduced varieties of plants (35 U.S.C. 161, 171, 173).

11. Heretofore only the inventor himself could file for protection except where the inventor had died, ws legally incapacitated, recalcitrant or could not be found and, in such instances, his legal representative and the owner of the invention, respectively, then had the power to file (35 U.S.C. 117, 118).

12. The Patent Section of the American Bar Association drafted complete legislation which was introduced in the House by Congressman Poff (H.R. 13, 951) and the Senate by Senator Dicksen (S. 2,597); copies of all legislative bills usually can be secured by writing one's Congressman or Senator.

13. The new position taken by the Administration has been indicated by an unpublished draft of a new bill limitedly circularized by the Patent Office in March , 1968

14. However, the use of the term "interference" will be dropped in favor of the term "priority."

15. This change will effectively eliminate the present 35 U.S.C. 104 which prohibits one from relying on acts in a foreign country.

16. Heretofore an applicant, in the absence of misappropriation of trade secrets or breach of a confidential relationship, had no actionable recourse against an infringer or claim for damage until his patent issued.

17. Note note 8, section (a) above.

18. It has been estimated that the average cost to each side in a patent law suit is $50,000 through the District Court level. This figure would not include the further cost of an appeal to the Circuit Court of Appeals

or to the Supreme Court.

19. The courts have the power to declare a patent invalid if such an issue is raised as a defense in an infringement suit or as an affirmative remedy in a Declaratory Judgment suit.

VII. How To Make Money Out Of Your Invention

Inventions may be sold to a manufacturer in various stages of development—when the invention is untried, searched through the patent office, patent applied for or patented. Because a patent is only as good as its weakest claim, many companies will purchase a searched invention and put their best attorneys to the task of obtaining maximum protection.

Items That Cannot be Patented

A nonpatentable idea, other than business selling techniques that can be easily copied by competitors once introduced, can be bought and sold following the ruling of establishing a contractural relationship. Remember, the nonpatentable idea will be worth much more if it is in definite concrete form and shown to be workable by models or samples.

There is an advantage to a manufacturer of being first in the field—often as great an advantage as a patent. Games and toy manufacturers hardly ever wait for the issue of a patent. Rather they make a patent search, possibly file a patent application, introduce the product, saturate the market, and then when sales fall off or when competition copies the product, they move on to still another product. Also, many companies go to the expired patents for sources of ideas and reintroduce items with no royalty payment necessary.

Getting More For Your Idea

If you cannot have a model made or test market your invention in-

The authors wish to thank the U. S. Small Business Administration for permission to edit and publish portions of this chapter.

stead of sending raw ideas to a manufacturer, have a patent search made by a competent searcher and send this and the patents he finds to the company. Try to get the searcher to write on his letterhead, stating that in his opinion the invention does not conflict with any known U. S. patent. The $50 spent for the search will aid the companies in consideration of your invention. Remember, first set up the contractual relationship by obtaining the company's disclosure form and written interest in being willing to evaluate your invention.

Try to sell a completed program, not just an idea for a product. Carry the idea through probable market testing and selling on paper. Get other peoples' reactions and comments to aid you in submitting a more complete presentation.

Usually, in order to receive compensation, the idea must be novel (patentable) and submitted in a clear concise form so that anyone skilled in the art could perform the objectives of the invention. Just an idea for the need of an invention to do such and such is not what manufacturers pay for, rather they want a workable invention that will perform as claimed.

By all means, protect your valuable inventions by patenting them. The money spent doing so will increase their worth.

$1000 per Invention

Many large companies state in their disclosure form that in the absence of a patent in most cases all they will pay is $1000 per invention. To the prolific inventor who has more inventions than he can ever market himself, the sale of six or eight inventions per year adds to his regular income. It is easier to sell the same company a second invention once you know the people and they know you and for possibly a higher figure than the thousand dollars. Thus you may have to sacrifice some money at the sale of your first invention to get an "in." Also, you will gain inside knowledge on what specific invention a company may be looking for.

Many companies offer tours of their plants. Take these—and inquire of the person conducting the tour of specific needed inventions the company may be looking for. Later, write the company to confirm the need. A few guides may not have the whole picture and could send you off in the wrong direction. A tour of an aluminum reduction plant sparked six needed inventions for one of the authors, which at this writing are patent pending.

A Piece of Advice

If your patent search indicates that at best your patent will bring about only a slight improvement and that the same improvement worked out in a slightly different way is found in a number of other patents in the field, it is suggested that you forget about the invention and turn your endeavors to other inventions. Companies will buy from freelance inventors when their inventions are big. These big inventions could increase the company's business.

There is an advantage in offering your invention to a large company when only a large company would have the resources to handle it. The medium or small company, because of its greater flexibility, may be the inventor's best bet for the future. Many of these small companies have a real problem locating new products to manufacture, while the big companies are turning down thousands of inventions.

Patent Pending and Patented Inventions

Many inventions are sold in the patent pending stage and royalties gained whether or not the item ever receives a patent. Only about 10 percent of companies will require an issued patent before they will consider an invention. The rest will evaluate at all stages of development.

How to Find Companies

A good company prospect list is to be found in checking the products going into the primary market in which you want to sell. For example, co-author Fenner wanted to sell his Permanent Press Solution, which would enable housewives to treat clothes at home, to department, variety, and drugstores. Ten companies that sell to these outlets were found by checking the company name on products found in these stores. The street addresses of these companies were found in the Thomas Register. All ten were interested in looking at the invention, four asked for additional information and samples, and one market tested the invention.

Go after companies that sell in the product market, not necessarily those equipped to manufacture your invention.

Statistically, pick about 85 percent of companies engaged in a direct line of manufacture or distribution to which the invention pertains.

The other 15 percent could be companies that are equipped to manufacture the invention.

The Thomas Register

The Thomas Register will be mentioned many times in this book and rightly so as their reference volumes contain an unlimited number of prospects for selling your invention.

Check the company's advertisements and what they are producing. Include those manufacturers that manufacture a component part of your invention. Also try some manufacturers not in the field.

Another feature of the Register is the financial rating of a company. A company in the red may want your invention very badly but if it goes into bankruptcy your invention may go unused and under court litigation for many years.

Your banker is another source for evaluating the financial stability of the firm that may be interested in purchasing your invention.

Use a Trade Journal

Consult a trade journal that is in the same field as your invention and check some of the advertisements to make your list of probable companies that may be interested in purchasing your invention.

To Sell Your Invention—Try Local Businesses

Businesses in your home town may be excellent prospects as you are readily available for consultation at no travel expense. Use your local or regional newspaper to circulate news about your invention and how you are seeking a local industry to sell the invention. Follow-up with a classified ad. The Chamber of Commerce or your banker or lawyer may be excellent sources of help to place you in contact with local businessmen.

Try to Get Publicity

Your local TV station and/or newspaper are always looking for news items, and an invention, particularly one that is unusual and potentially big, will interest them. Try, if possible, to get on a night time adult TV program put on by the bigger networks. A letter and a sample or

picture of your invention may land you an offer to appear on the show
and demonstrate your brainchild.

As mentioned in Chapter IV on Test Marketing, a well written letter
to the editor of a hundred better known popular or trade magazines
together with a description of your new invention and where possible
a commercial photograph will, it is estimated, generate in most cases,
approximately $3000 of free publicity.

Expose your patent pending invention to as many people as possible
to find the *right* person.

Patent Gazette

Use the listing services of the Patent Gazette under Patents Available For Licensing or Sale to publicize the availability of your patent to
industry. The cost is very nominal and the circulation large.

Use the names of companies inventing in the same class and subclass as your invention as a unique, often untapped source of company
prospects.

Is It Ethical?

The question arises as to whether it is ethical to submit the invention to more than one manufacturer at a time. Most manufacturers
realize that this is necessary in order to uncover the few companies
that would purchase the invention. A company may become angry if
you string it along for months and finally sell to another. Having two
companies interested in the same invention can sometimes increase
the worth of the invention in the eyes of both companies.

The inventor can get himself in a bind if he sends out his only model
and another bigger manufacturer asks to see it. But this is rare—send
a model of your invention if requested to do so by the first company
replying that has a good financial rating.

Selling Your Invention to the U. S. Government

The inventor desirous of selling his invention to the government
should write a cover letter describing in general terms the objectives
and novel features of his invention and submit this and a detailed description of his invention to the government agency he thinks would be
most likely to use it. The office of Invention and Innovation, National

Bureau of Standards, Washington, D. C. 20234, or the U. S. Small Business Administration could recommend the government agency that would be interested.

Some agencies may require that you sign a disclosure agreement before your invention will be evaluated; however, the Government does not normally require the use of a special form for this purpose. Your disclosure should contain:

1. A statement of the advantages of your invention that make it superior to similar devices in use or available.
2. Complete information on the method or principle used in your invention.
3. A description of your invention that is sufficiently clear to enable competent technical personnel to understand fully its construc-construction and step-by-step operation.
4. Any drawings, diagrams, or photographs necessary to disclose the true nature of your invention.
5. Any theoretical or actual performance data you have to show the operability and superiority of your concept.

Will the Government provide me with funds for the development of my invention? In some instances government agencies fund the development of specific inventive concepts. Usually the funding is made available on the basis of information provided in a research and development contract proposal. If you are interested, write to potentially interested agencies for instructions for submitting proposals. The publication. *Small Business Guide to Research and Development Opportunities*, containing general information on Government research and development activities, is available from the Small Business Administration.

Pitfalls in Contacting a Buyer

Before an inventor can sell his patent or license others to use it, he has to have a buyer—someone who is willing to assume the risks involved in an untried idea. In making contact with a buyer, there are certain pitfalls the inventor needs to be aware of.

For example, if the patent *has not yet been issued* on the invention (or if the invention has not yet been made public by marketing the product), the inventor or official of the small business concern may encounter reluctance on the part of a larger existing company. That larger company may not want to even accept information on the invention for licensing purposes.

One reason for this attitude is that in the free competitive economy existing in the United States, anyone is free to manufacture and sell, for profit, any product or device coming to his attention in the ordinary course of business, unless the product is covered by valid patent rights.

Substantial manufacturing concerns do not wish to handicap themselves by receiving information on what the inventor says is a new improvement or idea. Before receiving such information, these companies want to arrive at a clear-cut understanding as to their obligations to the person submitting the information.

Another reason is that existing manufacturers in a particular field may themselves have been working along similar lines. Under such circumstances, they do not want to assume undefined obligations, upon receiving information on what is asserted to be a new invention or improvement in their particular field.

Some companies have been held liable (in past court actions) for damages for using information on inventions obtained under a confidential or trust relationship not clearly defined when the information is submitted. For this reason, most large manufacturers have adopted a policy of not reviewing technical information in the absence of some form of a release contract from the person submitting the information. This contract usually provides that the reviewer of the information has no obligations except those imposed by valid patent rights.

It is difficult to give a formula or guide that will help all inventors (or small companies that are trying to sell an invention) in all situations of this kind. However, you can be aware of these pitfalls and seek the advice of experienced legal counsel when you are trying to establish a negotiating relationship with a buyer.

In addition to a sale of your invention, there are various types of license arrangements.

Suppose that as a buyer or seller you have solved the problem of initial contact for negotiations. What types of licensing arrangements can you consider? For one thing, they should be such as to provide a fair benefit or return for both of the parties to the contract.

Many factors need to be considered in arriving at the amount of royalty which the inventor or the company representing him should receive. Among them are:

1. The degree of novelty present in the invention, over what had been known before;

2.   The extent to which it may supplant similar products on the market;

3.   The scope of the claims of the patent, if a patent has been issued, or of claims likely to be obtained, if the patent application is still pending;

4.   Is the invention basic and pioneering in the sense that it is an entirely new product or technique?   Or is it merely an improvement on existing devices or techniques and perhaps subject to earlier issued patents of a more basic nature?

5.   The expenses the licensee must bear in placing the invention on the market;

6.   The profits anticipated by the licensee.   The most important consideration is to arrive at an understanding likely to benefit mutually both parties to the agreement.   While the inventor or the small business representing him will seek a substantial royalty, there undoubtedly have been many instances where an invention has not successfully reached the market because too high a royalty or other payment was demanded.   This high payment was demanded without due consideration of the contribution that must be made by the licensee in setting up manufacturing facilities and finding and reaching the market where the demand for the product exists.

Exclusive or Nonexclusive Licenses

The owner of the patent rights may, if he so desires, grant an *exclusive* license which permits *only one* company to manufacture and sell the product (or to practice the technique covered by the patent).   In such instances, it may be reasonable to ask for a higher royalty payment.

It may also be reasonable to seek a substantial downpayment upon granting an *exclusive* license.   Because the exploitation of the patent rights are being put in the hands of a single company, it may also be reasonable to ask that the licensee should agree to pay a minimum annual royalty, throughout the term of the license, in consideration for the exclusive rights granted.

It is not possible to suggest what royalties or payment would be reasonable.   These amounts will depend on the importance of the invention, the scope of the patent, and other factors existing in the particular industry.   And they will need consideration in each case.

It is also possible to grant more than one license, under a patent. In fact, any number of *nonexclusive* licenses may be granted.   In general, royalty for a nonexclusive license would be less than that applicable for an exclusive license. This is true because the licensee would be competing with others authorized to manufacture and sell the same product.

Reservations in the License

A license may be granted for a certain part of the United States only. For instance, it may be limited to a certain state or a group of states.

It may be limited to a specified field of use. For example, if the invention is concerned with a motion picture projecting machine, a particular license could be limited to the manufacture and sale of machines designed for home as distinguished from theater use. A license might be limited to the right to use the invention, thus not granting the right to manufacture and sell.

A point to keep in mind is that the owner of a patent enjoys considerable flexibility in the number and types of licenses he may grant.

Other Obligations of Licensor and Licensee

The licensor may or may not assume a positive obligation to enforce the patent against infringers. However, it is obvious that a licensee might reasonably expect he would not be placed in an inferior competitive situation by being required to pay royalties while others can manufacture and sell the same product without paying royalties.

In the case of an exclusive license, the licensee may agree to assume at least some of the responsibility for enforcing the patent against infringers. On the other hand, the nonexclusive licensee might not normally expect to bear any of these responsibilities.

It may be desirable to provide that the licensee will exert reasonable efforts to manufacture and sell the product in accordance with the patent rights licensed. This is particularly true in the case of an exclusive license. On the other hand, agreements that prohibit a licensee from engaging in the business in similar products that fall outside the scope of the claims of the licensed patent are not legal.

Right to Terminate The License

The term of a license is usually for the life of the patent, but may be for a lesser term. The right to terminate a license short of the full term granted may be important both to the licensor and to the licensee.

The licensor may need this right, particularly if he is granting an exclusive license and if no initial payment or minimum annual royalties are provided for. On the other hand, the licensee may consider the right to terminate a license to be important, particularly where other companies may infringe the patent by manufacturing and selling without payment of royalty.

Selling Rights Abroad

Inventors may want to consider the possibility of selling or licensing patent rights in foreign countries. Because foreign countries have their own patent laws, it may be desirable to establish patent rights in those foreign countries that may be likely markets for your invention.

The U. S. government has established treaty relationships with other nations whereby inventions made by United States citizens can be protected abroad. That is, they can be protected abroad if appropriate and timely action is taken to secure foreign patents corresponding to the United States patent rights. This insures a bargaining position for United States enterprises in dealing with foreign companies.

SBA Information and Patent Rights

In its function of assisting and encouraging small business, the Small Business Administration provides information and services in the area of patents and new products.

First, SBA helps small business concerns to obtain nonconfidential research and development data. Often it is not readily obtainable by the owner-manager. He does not have the time to check many sources or he does not know where to go.

Second, when a small business asks for help on problems involving new products, SBA obtains recommended solutions from sources in industry and Government for the inquirer.

What is an Untried Idea Worth?

Often it is difficult to determine what a new invention is worth. The value of anything depends, to a great extent, on who is talking—on whether you are buying or selling.

This fact is especially true when a small company and an independent inventor are discussing the buying and selling of a patent. The patent and the invention it covers mean certain things to the inventor. To the small company it means other things—opportunities for expansion, as well as problems.

First, suppose you are the inventor of a new item. You are proud of it because you have created it. You have developed the invention out of your experience and knowledge. It is the product of your creative ability.

In selling your patent you want compensation for the work you have done in developing your idea as well as for the creative effort involved. You can measure your work—so many hours or days or months. However, creative ability poses a difficult problem. Can anyone say how much it is worth?

Now, imagine that you are sitting in the chair of the owner-manager of a small company that is considering buying a patent. Look for a few moments at his opportunities and problems in connection with a new and untried idea or invention.

"I can get into a new line with this invention," you think. Or "I can increase my company's sales by using this invention to improve my present product." But then you say to yourself, "that is, if everything goes all right."

Then as owner-manager you begin to consider some of the problems connected with buying the invention. First, with the payment to the inventor for this patent, your financial requirements are just starting. You will have several other major expenditures in connection with refining the invention into a salable product.

You will be spending for product engineering, production engineering, and market research. How much for each is another problem.

The point is that as owner-manager you will have to spend this money to get the new invention ready for production. As you begin to make the new product, you will need to spend still additional money. You have to pay for raw materials, for labor, and the other manufacturing items necessary to create an inventory.

Finally, you will have to tie up additional working capital to finance the accounts receivable you hope will result from the sale of the product.

You may have a good idea of how much you will need to spend and of how well the new product may sell. But as you discuss the purchase of the invention with the inventor, you do not know whether the new product will be successful.

For example, you could spend thousands of dollars getting the new invention to market only to see it fail because you had misjudged the market. Or it might fail to sell because of other conditions over which you had little or no control. You are taking risks.

These risks and problems and expenditures will, to a large extent, determine what the new invention is worth to your small company. The question you as owner-manager have to answer is: How much can I spend on this invention and come out with a product that I can sell at a profit?

Briefly, these are the major aspects of the economic side of buying and selling a patent. Perhaps, at times, they are more complicated than the legal aspects of buying and selling a patent.

Conditions in the market place often do not lend themselves to precise definition. Perhaps their main characteristic is risk-risks that both buyer and seller must accept as part of a free economy.

Such risks must be considered by both buyer and seller. Both should be willing to negotiate in a spirit that recognizes risks and the necessity of sometimes compromising in order to overcome them.

The best advice to give inventors on how much an invention is worth is not to be too demanding. Often times inventors pick figures out of the air with no realistic value as to the worth of the invention to the prospective company. Shop around for offers but do not let too much time lapse if there is only one company interested. It is far better to sell the invention for a cash payment and royalty license than for a single larger cash settlement. Make certain there is a nonshelving clause and a minimum yearly payment for the use of the invention in any contract you sign.

A royalty rate of 2 to 5 percent of the manufacturer's selling price is common if the article will be mass produced, however, if the item is expensive and of a limited market, a higher figure is advisable. Your experience in test marketing will be of great value to you in establishing a royalty as you will have an idea of manufacturing costs and what the market will pay for the invention.

How to Protect Salable Ideas

Marketable ideas

Many new ideas are basically intellectual property. But if put into practice they could be profitable both to the originator and to the busi-

nessman who adopts them. Improvement in the design or in the appearance of an item already patented or available to the public falls in this
class. Also, the redesign of an article popular years ago to make it
acceptable again may prove profitable. Likewise, the substitution of
superior materials for inferior ones may enhance the salability of a
product.

Ideas for improving various compositions, including formulas, by the
mere addition of a new flavoring, a new scent, or the like, may make
it possible to offer a superior product to the public, and thereby increase
profits. The important questions are: How can the idea man or businessman promote these valuable unpatentable ideas? How can they be
protected? What steps can be taken to profit from them?

Growth of legal opinion

As far back as the early English courts, it was held that a man had
title to his originations, but unless protectable by established law those
ideas became public property as soon as divulged. This old common
law has been passed down to become part of our present code. Many
idea men, therefore, are led to believe that their ideas cannot be
protected in any way. As a result, numerous potentially profitable
concepts have been abandoned. There is little, if any, actual statute
law governing the exchange of ideas, but a large body of court
decisions in this field has accumulated. Much can be learned from
the more important cases.

Cases in point

The requirement that the possessor of an idea must be able to prove
that it actually originated with him was brought out by the suit of
Moore v. the Ford Motor Company, 43 Fed, (2d) 685. Moore had conceived a thrift purchase plan adapted for use in the sale of automobiles.
He wrote to the Ford Company that he would like an opportunity to
submit the plan, which he believed would increase the sale of their automobiles. Ford replied:

> If you will kindly write us in detail regarding the plan you have in mind for
> increasing the sale of Ford cars, understanding that in doing so there will be
> no obligation on our part, we will be very glad to give the matter our careful
> attention and advise you whether or not we would be interested in the plan.

Thereupon Moore submitted his plan in a letter, concluding with
the following paragraph:

The above is a general idea of what I have in mind.   I understand it is subject
to amendments and eliminatións, but if it is usable I would very much like to
aio in perfecting it.   However, as called for in your letter, I am writing you
with the understandinp that there is no obligation on your part.

The Ford Company subsequently returned Moore's letter stating
that it would not be interested in the proposition. At a later date,
Ford put into effect a weekly purchase plan that became nationally
known. That plan was similar to the one submitted by Moore except
as to differences in detail.

Moore then brought suit, contending that his plan had been appropriated by Ford. In the trial, Moore relied on his own evidence—
namely, the correspondence between himself and the Ford Company,
plus the copy of the plan as presented to Ford, which the latter had
returned, marked "not interested." Ford called in witnesses testify-
ing that various Ford dealers throughout the country had already been
using similar plans—that is, weekly payment schemes whereby terms of
purchase could be made easier.

The court held that there was no piracy because there were too many
differences in detail between Moore's plan and the one put into effect
by Ford. Furthermore, said the court, the basic idea appeared to have
been used in Christmas savings club plans, which were known throughout the country prior to Moore's proposal. Because the inventor could
not establish that he was the first and true originator, nor prove definitely that the Ford Motor Company had appropriated or copied his
idea, Moore did not have any ground on which the suit could be sustained.

The requirement for protecting an idea with a contract was brought
out in the case of Bowen v. Yankee Network, Inc., 46 Fed. Sup. 63. Bowen
contended that William Wrigley, Jr., Company pirated his valuable
and novel idea for a "radio presentation." Bowen had submitted the
plan to Wrigley, which after some delay returned it as unacceptable.
Wrigley later disclosed the idea to Yankee Network, Inc. Soon, a weekly radio presentation entitled "Spreading New England Fame," containing the features and ideas set forth in Bowen's proposal, was produced on the network.

In court it was brought out (1) that Bowen voluntarily submitted his
idea to Wrigley; (2) that because of the voluntary submission there
was no breach of trust or contract; and (3) that there was no correspondence or other evidence to show that the disclosure of the idea to
Wrigley had been done with the understanding that there was any limi-

tation upon the use of it by the company.   As a result, it was held by the
court that Bowen could have protected his idea by contract, but that
he failed to do so when he voluntarily communicated it.   Whatever
interest he had in the idea, therefore became public property.

From this case we learn: (1) An idea made public, either by word of
mouth or in writing, immediately becomes common property, and unless the plan is revealed under contract or by confidential disclosure,
anyone can make use of that property without infringing any rights;
(2) the voluntary submission of an idea does not set up a contractual
relationship between the originator and the other party; (3) because
of the lack of contract, no action of any kind can be brought by the
originator for breach of trust or contract; (4) ideas can be protected,
provided the originator follows certain procedures governed by the
law of contracts.

Another decision that emphasized the importance of drawing up a
contract to protect an idea was rendered in the case *Equitable Life Insurance Company*, 132 N. Y. 265.   In this suit it was held that, without
denying there may be property rights in an idea, trade secret, or system, it is obvious that its originator must *himself* protect it from escape
or disclosure.   If the innovation cannot be sold or negotiated for or
used without a disclosure, it would seem proper that some contract,
either expressed or implied, should guard or regulate the divulgence.
Otherwise, the idea becomes the acquisition of whomever receives it.

The law of contracts applies in *all* such instances, but for it to be
binding there must be a definite "meeting of the minds" (i.e. agreement
among the parties concerned).   If you are a businessman looking for
suggestions, by very careful if you advertise for them.   You should
state that ideas to be considered must be original and novel, the conception of the person submitting them, and that they must be submitted in explicit form.   Any ambiguity could easily result in nullifica-
tion of a contract because a definite proposal had not been made.

Furthermore, the fact that you as a businessman consent to examine
an idea does not mean that you have to buy it or put it to use.   Therefore, in your contracts or correspondence you should insert a proviso
whereby you do not obligate yourself in any way by appraising the idea
to be submitted.   This, of course, does not free you from obligation
in the even that you *do* make use of the idea submitted.   In the *Moore
v. Ford Motor Company* case, it was held that all danger of subsequent
demands or obligations cannot be avoided by merely stating "no obligation whatsoever."   You should, therefore, protect yourself against

any obligation to *utilize* the plan.

Some businessmen protect themselves by setting up special methods for handling ideas. Frequently they are channeled to one individual, making it impossible for the information contained in that idea to become accessible to employees at large. By doing this it can be proved more easily in court that the idea has not been used or permitted to escape.

The acceptance of conditions by the signing of an agreement constitutes a "meeting of the minds," and once that has been accomplished, a definite contract has been established. The businessman is bound not to divulge any features not already known to him or to others, or to make use of the idea without proper compensation to the inventor. Likewise, the businessman is protected against any unwarranted claims.

When the businessman agrees to review the invention and report on his decision he also understands that he will receive complete details —something on which a definite opinion can be based. Thus, in order to make the contract binding and valid, it is up to the idea man to furnish all pertinent facts.

After the originator of the idea has submitted his material to a businessman under a definite contract, it is up to the latter to fulfill his obligation by carefully examining the material and reporting his findings to the originator within a reasonable time. In his reply to the originator, the businessman should state whether or not he is interested in the actual purchase and use of the idea.

Checklist for Handling Unpatentable Ideas

The following checklist is offered governing submission and acceptance of profitable ideas:

The idea man can:

1. Establish priority to an idea.
2. Protect his idea by complying with the law of contracts.
3. Safely deal with others so that they may examine his idea to find out if they want to buy it.
4. Sue for breach of contract following misappropriation of an idea submitted in confidence and under contract.

The idea man cannot:

1. Broadcast the idea and still retain exclusive control.

2. Recover damages if the idea or any of its details are submitted unsolicited.
3. Recover damages for misappropriation of his idea unless he can prove that he is the first and true originator.
4. Recover damages unless there is a violation of a definite "meeting of the minds" between him and the other party as to specific conditions of disclosure.
5. Recover damages unless there·has been obvious copying or use of any or all of the submitted idea.

Business managers should:

1. Be willing to consider ideas of originators outside their business employment.
2. Require submission of complete details under specific agreement.
3. Report promptly their reactions and possible interest.
4. Retain all information confidentially.
5. Deal fairly with the originator.

Business managers should not:

1. Review any material or idea *not* submitted under agreement.
2. Divulge any information on an idea received.
3. Make use of any features of an idea (even when submitted at their own request) without express permission of the originator.

Thus, an idea man who creates something worthwhile, even though no patent or copyright protection is available, may be able to cash in on his innovation. Legal decisions aid him if he follows the procedure by which many unpatentable ideas can be safely handled. When offering intellectual property, he must be able to prove that he is the first and true originator, and he must have the assurance of a contract that his disclosure will be treated confidentially. Once an idea leaves the mind of the originator, except under specific contractual arrangement, that idea becomes public property. A business manager must deal fairly with originators of ideas who are willing to release their conceptions for consideration as to possible use. Likewise they must protect themselves against suits for misappropriation of new ideas.

Companies in general would prefer the inventor not give the date of conception or filing so that they cannot be accused of trying to fabricate records in an attempt to prove that they were already working on such an invention. The knowledge of your conception and filing dates would be valuable to a manufacturer in a patent interference or suit.

Because of the necessity to communicate the disclosure to people to determine its true potential, companies cannot accept disclosures of inventions in confidence. In general, it is believed safe to submit inventions in accordance with manufacturer's requirements providing a contractural relationship is established.

Fill out and date both copies of a company's disclosure forms. If only one is sent, fill it out and machine copy the document. This can prove to be very valuable to the establishment of a date years later.

The following is a composite disclosure form from six companies and hence contains a maximum number of conditions that illustrate extremes and not usually found in a typical form.

1. The XXX Company is willing to consider any suggestion that may be made, but does so only at the request of the person who has the suggestion.

2. No obligation of any kind is assumed by nor may be implied against the XXX Company unless or until a formal written contract has been entered into, and then the obligation shall be only such as is expressed in the formal written contract.

3. I do not hereby give the XXX Company any rights under any patents I now have or may later obtain covering my suggestion, but I do hereby, in consideration of its examining my suggestion, release it from liability in connection with my suggestion or liability because of use of any portion thereof, except such liability as may accrue under valid patents now or hereafter issued, or under any formal written contract.

4. XXX reserves the right to require at any time that all submissions be in or be translated into the English language.

5. XXX is not obligated in any way to use any of said ideas or information. However, the use to be made by XXX of any ideas and/or information submitted to XXX and/or any of its representatives, and the amount of compensation, if any, to be paid for such use are matters resting solely within the discretion of XXX.

6. XXX is not obligated in any way to discuss or to give any reasons for its decisison respecting any submission.

7. XXX assumes no obligation of any kind to compensate the submitter (and/or his principals, if any) for any costs or expenses incurred in connection with the submission of the ideas or information.

8. XXX is under no obligation of any kind to return any material submitted to XXX but XXX has the right to make and retain a copy or copies of any such material.

9. In the case of an idea that has not been patented and on which no patent application is pending, the company must be the sole judge of the value of the idea, which shall not exceed $1,000 unless for some extraordinary reason the company voluntarily elects to pay a larger amount.

The undersigned hereby represents and warrants that he has the right to disclose the ideas and/or information in question; that he has the right to make the present agreement; that there are no outstanding agreements of any kind which are inconsistent herewith; and that the submission in question was not solicited by XXX and/or any of its representatives.

As an inducement to XXX to consider the ideas and/or information the undersigned agrees that the submission thereof (as well as any additional material which may be hereafter be submitted as incidental to the material originally submitted), and any consideration which may be given to them at his request by XXX shall be in accordance with the above terms and conditions.

I accept these conditions and, accordingly, request you to consider my suggestion which relates to_____
_____.

ACCEPTED THIS_____DAY

OF_____A.D., 19___

_____
        (Signature)

_____
        Address

WITNESS:_____   _____
                                           Date

WITNESS:_____   _____
                                           Date

Presenting the Invention

It is important in your preliminary material to be brief and to the point. Only submit one invention at a time. Allow each invention receive to singular attention.

Encourage manufacturers to call you about any questions they may have. Give them a phone number where you can be reached during the business day.

The cover letter should be carefully worked out and list possible com-commercial merits. Do not burden the reader with too much detail. Save this for the description. Do not use high-pressure sales techniques as this will disgust the reader.

Get the attention of the reader of your cover letter by explaining the invention, making it perfectly clear and understandable to the executive making the evaluation. It is important to get the idea across without taking up too much of the reader's time.

The data gained in market testing, for example, production costs, working blueprints or patterns that would allow a manufacturer to go right into production without delay or experimentation, equipment, materials to be used, etc. can be important selling points.

You might want to dress up your presentation by placing all of your papers in an attractive folder. This would make it easier to handle in going to the various departments necessary to review your invention.

In addition to a well written cover letter, you should enclose a description of your invention and a drawing showing its mode of operation. Helpful in the description is a list of all the features that the invention has—either new or improvements over competitive products. Many inventors omit the title until last because with the list of features before them a more descriptive title is often possible. Do not make the mistake of naming the invention with your family name, such as, "Smith's Wonder Worker."

You can probably make a drawing of your invention with ordinary instruments—pencil, ruler, compass—that will illustrate the features and method of operation. Number each part, keeping the same number of the part in subsequent figures. Refer to the part by number in the description. It is recommended to use heavy typewriter paper gluing or taping the sheets together if necessary. Copy your drawings on machines presently available in public buildings. These copies often look better than the original.

Good drawings are essential. Where possible, keep the illustrations and description together on the front of a single sheet of paper to keep from having to turn the paper during the reading.

Write and rewrite your description and ask someone to read and explain it back to you. Do they get the whole picture? If the invention is complicated, it may be necessary to ask a patent attorney or engineer to write up a description for you. He will know the correct technical terms and language that should be employed.

You may wish to copy your drawings—the open top electrostatic copiers are excellent.   Then you can pencil in the description and later type it on another copy or the original.   It may be necessary to retype the description to make it fit into the final arrangement without being too crowded.   Photo offset printing of your layout will give you an attractive brochure.

A No. 2 lead pencil will often be easier to handle than Indian ink. It may be necessary to have several views and elevations.   Money invested in drawings by a draftsman or architect for a complicated invention will be well worth it in making the presentation attractive and fully clear.   You may want to keep in mind drawings that not only make the device clear but attractive as well.   These drawings, if patent office style is followed on Bristol Board, could be used for your patent application.

The Importance of a Model

A model that works, no matter how crude it is, is an invaluable aid in selling an invention.   The list in the Appendix will be of value in selecting the model maker you want.   If you need further names, the telephone directory of Chicago or New York, or the Thomas Register, could supply you additional names, addresses, specialties of hundreds of model makers.

Be certain to explain every detail to a manufacturer on how it should be operated to avoid damage, should it be left with or sent to a manufacturer.

A Radically New Invention

Negative replies are often received with a radically new invention and the inventor should not be discouraged.   Unfortunately, a lot of new inventions are reviewed with skepticism and by people who know exactly why the invention will not work.

How Much Does It Cost

The following is a breadown of expense incurred by Associated Ideas to promote its fuel cell invention.

100 Form Letters to determine interest of manufacturers to obtain their disclosure form and to establish contractual relationship. We used a standard mimeographed form letter as shown below.	$ 12.00
Answers to form letters, completing disclosure form and enclosing a one-page cover letter explaining the advantages of the fuel cell system.	25.00
Enclosed copy of fuel cell patent application consisting of 8 pages of type (mimeographed) and 3 pages of drawings (offset printed). Fifteen extra copies had to be ordered to place a minimum printing order. One copy was used as the patent application.	50.00
Mimeographed form letter to determine why the company was not interested in the invention (opposite).	1.00
Follow up letters giving more information.	10.00
Long distance phone call.	5.25
	$103.25

Date

Dear Sir:

I have an invention that I believe may be of interest to your company. Would you please inform me of the procedure by which I may disclose this invention.

Sincerely,

Terrence W. Fenner
4447 Lafaye Street
New Orleans, La. 70122

We Profit By Our Mistakes

You are asked to evaluate_____
for possible purchase by your company. A polite negative reply is very
disheartening to an inventor when possibly the invention is of value
but does not fit into the manufacturing or marketing plans to the firm
submitted. If the invention offers no improvement over the existing
art, it is better that the inventor be informed so that he can channel his
efforts into other more productive ventures.
If a negative reply is forthcoming on your evaluation of this invention,
would you be so kind as to comment on why the invention was not
accepted. This is greatly appreciated.

_____The invention is good but does not fit into our immediate
plans.

_____The idea is good but the invention needs improvement in
the area of_____

_____

_____The invention offers no improvement over the existing art.

_____The invention just doesn't work.

_____The invention is not new.

Remarks:_____

_____

_____

An Improved Fuel Cell System and Method of Operation Thereof

ABSTRACT OF THE DISCLOSURE

This invention is designed to greatly improve the existing art of fuel cells by replacing rigidly solid catalyst electrodes with a moveable "slush" medium which can be stirred up, removed from the electrode compartment, reactivated and reintroduced. While removed from the reactive area of the fuel cell, the electrode medium would be pressurized with fuel and oxidant, if desired, for improved amperage. The existing art is also improved using battery ingredients in "slush" form for a full cell and the whole system is made more useful through a coin-operated electric outlet device for recharging the "slush" fuel cell.

CROSS REFERENCES TO RELATED APPLICATIONS: NONE

BACKGROUND OF THE INVENTION

1. Field of the Invention

    This invention relates to the field of fuel cells.

2. Description of the Prior Art

    Present day fuel cell systems suffer the disadvantages associated with solid electrodes, usually carbon blocks or grids coated with platinum or palladium, or pressed or sintered catalyst electrodes; however, these electrodes become deactivated with use either through catalyst metal poisoning by the fuel or electrode polarization or physical removal of the catalyst material by abrasive action with use. It would be very useful, therefore, to have a system by which the electrode's catalytic material can be removed, reactivated and reintroduced and at the same time produce greater amperage.

SUMMARY OF THE INVENTION

The object of this invention is just such a system to overcome the above problems wherein the electrode medium is liquid, such as the catalytic metals or their salts in a molten state capable of reacting to produce electricity or the catalyst metals in particle form, or the catalyst metals coated onto conductive particles with the resultant particles suspended in a conductive liquid or other suspending material, or in a quasi liquid such as in the form of a moveable powder. Because this material is not rigidly solid block but rather liquid or quasi liquid, it can be reacted, circulated out of the reactive area, reactivated and reintroduced by pumping, gravity flow or other means.

Another part of this invention is a fuel cell having a liquid or quasi liquid medium capable of being reacted to produce electricity and capable of being circulated out of the reactive area, externally saturated or

chemisorbed with hydrogen, methane, methanol and like type fuels and
separately saturated with oxygen, nitric acid and like type oxidants
followed by feeding these saturated chemisorbed catalyst mediums back
into the fuel cell to produce electricity of a voltage and amperage proportional to the fuel system selected, its saturation or chemisorption
into the catalyst of each electrode medium and the flow of said saturated
medium back into the cell. After discharge the electrode medium would
continue the cycle of being resaturated and reintroduced.

The advantages of this system are many. The catalyst containing medium
can be chemisorbed and held ready awaiting use and the amperage can be
regulated by the flow of absorbed (activated) catalytic material into the
cell to react. The circulation would also permit the catalyst to be
regenerated from time to time by chemical activation, burning off the
impurities, or fresh catalyst added. Another advantage to this system
is that the catalyst electrode medium would be pressurized externally to
a relatively high over pressure of fuel gas, if desired.

Still another part of the invention would be an improved fuel cell continuously fed the ingredients of a battery such as lead, lead dioxide,
sulfuric acid, manganese, silver, zinc, cadmium, nickel, etc., in elemental
form or as their salts, when and only when powder is needed. The ingredients
would be fed into their respective electrode compartment containing a porous
conductive separator with only the electrolyte, ions and electrons having
the mobility to travel freely throughout the system.

Batteries formed by metals in elemental form being converted to their oxides
would likewise find use in this invention with either the oxygen being
supplied from the air or for greater reactivity of the system as pure oxygen
contained in a closed system, supplied for reaction and regenerated during
recharging.

A filter device in the system to remove the reactants for storage and
recharging would minimize the amount of electrolyte needed for the entire
system. A good name for describing the consistency of the electrode compartments medium is that of a slush.

Carbon or like type particles may be used as a surface to accept plating
on recharging the spent battery ingredients and would be an aid in keeping
recharged particle size small and regulated.

The ingredients would be cycled, stirred up, pumped in and out of the reactive
area, and reactivated when desired. Multiple cell systems would likewise
build up the power as is done with conventional batteries.

The electric automobile industry would probably be the largest user of a
fuel cell being fed the ingredients of a battery for propulsion. The
quantity of battery ingredients to drive 75 to 100 miles per day or there-
abouts would be self-contained in vessels of particles to react to produce
electricity and vessels to receive the spent discharged particles for
recharging. Reactivation, recharging, while parked and at night from

rectified house current would be one method to make the system economically attractive. Economics of $2.50 to $4.00 would operate a rechargeable fuel cell battery car propulsion system for a whole month.

Proper agitation during reactivation with a propeller will keep the reformed battery ingredient's particle size small and suspended and hence present no real problem with particles clogging up pipes.

One system for maintaining the proper amperage and voltage would be to have two two-compartment vessels for the battery ingredients, one set to deliver to the fuel cell new or reactivated battery ingredients and the other set of two vessels to store the spent ingredients to reactivate them at will. The flow to and from the vessels through the fuel cell is reversed after the first two vessels are discharged. All the vessels would have recharging means.

Some current would be wasted each day in a motor(s) to pump the ingredients and for propellers to keep the particles suspended but the overall efficiency of this rechargeable system far outweighs these disadvantages.

A rinse system using fresh or filtered electrolyte would replace the reactive or spent particles when the battery would be shut off, if desired.

There are many advantages for a fuel cell system being fed the ingredients of a battery, the greatest being the indefinite rechargeability, the economics of operation, and electricity production on demand.

A further part of the invention is an improved fuel cell electrode made of electricity producing catalyst particles or catalyst coated particles closely packed yet still moveable or quasi liquid to comprise an electrode having a much greater surface active area from which to produce electricity.

Still another part of the invention is an improved coin operated and activated time limiting parking meter having an electric outlet to facilitate recharging of automobile fuel cells or storage batteries during a parking interval. Said recharging device would add many additional miles of driving to an electric car propulsion system making it a great deal more efficient.

Said improved coin operated parking meter and recharging outlet would be dual or single coin operated and activated for conventional gasoline driven and electric propulsion cars for parking and/or recharging. The current and amperage supplied from the outlet would be of a suitable type voltage and amperage to facilitate recharging. Stepped down rectified alternating current would probably be the most practical with the step down transformer and rectifier being self contained in each parking meter or used to feed a bank or street of improved parking meters with recharging outlets.

Still another part of the invention is a coin operated and activated time limiting electrical outlet device for recharging automobile fuel cells or storage batteries.

BRIEF DESCRIPTION OF THE DRAWINGS

Figure 1 illustrates a fuel cell having electrode compartment (10 and 11) containing electricity producing particles such as platinum, palladium, semiconductors like nickel sulfide, zinc oxide, nickel, nickel boride, platinum in elemental form or deposited on carbon particles and the like, suspended in an electrically conductive liquid which may or may not be the same as the electrolyte for the cell. (9). The fuel and oxident for the cell is supplied through pipes (3 and 4) into each electrode compartment, with means (5 and 6) used to circulate the electrode medium containing electricity producing particles out of the reactive area, reactivating the catalyst by burning, chemical or electrical cleaning, or other suitable means and reintroducing the electrode particles using pumping, gravity flow or other suitable means. Membranes (7 and 8) function as to contain the conductive medium and to allow electrical conductivity and ion mobility so as to give a charge to each electrode compartment which is removed by electrodes (1 and 2) or other suitable means.

The catalyst containing electrode mediums may be saturated with fuel while out of the reactive area and reintroduced for greater amperage and efficiency and would find use together with a normal fuel delivery system, if desired. It may be possible that in some cases the system would work better on just circulating the fuel electrode medium and having the other electrode compartments of a conventional nature.

Figure 2 illustrates a fuel cell having electrode compartments (5 and 9) containing electricity producing particles such as platinum, palladium, nickel, nickel boride, platinum in elemental form or deposited on carbon particles and the like suspended in an electrically conductive liquid which may or may not be the same as the electrolyte for the cell (6). The cell is equipped with means (3 and 4) to circulate the electrode medium containing the electricity producing particles into vessels (not shown) wherein the catalyst is saturated with fuel and oxidant respectively and then introduced back into the respective sections of the fuel cell by pumping, gravity flow or other suitable means to react in the fuel cell system to produce electricity proportioned to the fuel selected, its saturation concentration in the electrode medium and the flow rate back into the cell. After discharge the electrode mediums would be recycled out of the reactive area and resaturated (chemisorbed) with fuel and oxidant and reintroduced.

Membranes (7 and 8) function as to contain the conductive medium and to allow electrical conductivity and ion mobility so as to give a charge to each electrode compartment which is removed by electrodes (1 and 2) or other suitable means.

Figures 1 and 2 are not drawn to scale nor would the electrolyte compartment necessarily in use be as represented. In the figures the electrolyte compartment is distinct and expanded to show the separate nature of each electrode compartment. A single membrane may in practive suffice to separate one cell or electrode compartment from another. Multicell systems would find use in this invention to make the amperage obtainable a more useful entity.

Figure 3 illustrates a fuel cell having the ingredients of a battery such as lead, lead dioxide, sulfuric acid, silver, cadmium, nickel in elemental form or their salts, manganese, manganese dioxide, zinc and oxygen, and the like being supplied into each separate respective electrode compartment cell with the ingredients either being supplied in fresh virgin form or recharged previously and in a suspension in an electrically conductive medium which may or may not be the same as the electrolyte of the fuel cell and having suitable means (not shown) such as pumps to circulate the medium out of the reactive area and into a storage area for convenience of recharging these ingredients when desired.

Propellers (not shown) would be found in the fuel cell and in the recharging storage and dispensing vessels to keep the recharged particles small and in suspension. The storage and dispensing vessels would be equipped with recharging electrodes (not shown) which would be attached to rectified house current for recharging or other suitable recharging means.

Figure 3a simply shows an exchange of flow from the recharged particles in the storage vessel through the fuel cell and into the empty vessel for storage of the spent particles which would be recharged at will.

Figure 4 illustrates an improved fuel cell electrode compartment having catalyst particles or catalyst coated particles closely packed together to form a moveable quasi liquid having a greater surface area for improved electrical generating capacity.

Figure 5 illustrates an improved coin operated time limiting parking meter and recharging outlet for the recharging of automotive fuel cells or electric storage batteries.

Figure 6 illustrates coin operated time limiting electric outlet device for supplying electricity suitable for recharging automotive fuel cells or storage batteries.

DESCRIPTION OF THE PREFERRED EMBODIMENTS

The improved fuel cell catalyst electrode compartment would have liquid or free flowing solid particles in a liquid such as platinum, palladium, nickel, etc. in a suspension in the electrode compartments with the fuels, mainly gases, bubbling through these electrodes while not allowing the catalytic particles into other parts of the cell. This would be accomplished via a porous conductive separator with only the electrolyte ions and electrons having the mobility to travel freely throughout the system. The advantages of this system are that the electrode materials would be cycled, stirred up and pumped in and out and reactivated as desired. Multiple cell systems would likewise build up the power as is done with conventional batteries.

In certain cases, such as when one uses cylinder pressurized fuels, cryogenic liquid fuel or natural gas under pressure, etc., the energy of the expanding gases could be used to turn a turbine to pump the various mediums. This would be important for offshore installations and aids to navigation.

The electrolyte of the systems would be potassium hydroxide, or a solid polymer ion exchange membrane, or any of a host of electrolytes compatible with the fuels selected and the electrode compartment materials.

Colloidal catalyst metals for the electrode compartment can be made by the following methods:

1. Platinum metal is dissolved in a minimum amount of aqua rigia and the resulting solution diluted to the extent desired with water. Soda ash or other suitable basic material is then added to neutralize the acids present plus a slight excess. Hydrazine or other suitable reducing agent is added to effect the reduction of platinum to the colloidal form, which is stable as an aqueous suspension for an indefinitely long period of time.

2. The above procedure can be used to produce colloidal palladium.

3. Colloidal nickel is very difficult to prepare in aqueous media being very unstable. However, if nickel formate is finely dispersed in glycerin and the glycerin heated below its boiling point for a time, the nickel formate decomposes to yield metallic nickel which exists as a colloidal suspension in glycerine.

CLAIMS

We claim:

1. An improved fuel cell, with said improvement consisting entirely of having one or more electrode compartments containing a catalyst medium capable of being reacted upon to produce electricity, and capable of being circulated out of the reactive area, reactivated and reintroduced into said electricity producing reactive area.

2. The process of claim 1 wherein the electrode medium is platinum deposited on carbon particles.

3. The process of claim 1 wherein the electrode medium is a colloidal metal, that metal being selected from a class consisting of platinum, palladium, nickel or silver.

4. The process of claim 1 wherein the electrode medium is a finely powdered metal, that metal being selected from a class consisting of platinum, palladium, nickel or silver.

5. The process of claim 1 wherein the electrode medium is a finely powdered compound semiconductor such as nickel sulfide, nickel boride, zinc oxide, etc.

6. An improved fuel cell with said improvement consisting entirely of having one or more electrode compartments containing a catalyst medium capable of being reacted upon to produce electricity and capable of being circulated into and out of the reactive area, and capable of being

externally saturated with fuel and oxidant and fed back into the
reactive area of said fuel cell to produce an amount of electricity
proportionate to the fuel selected, the concentration of fuel saturation and chemisorption into said electrode medium and the flow rate
of said fuel saturated medium into said electricity producing reactive
area.

7. The process of claim 6 wherein the electrode medium is coloidal platinum
and the fuels selected for separate external saturation of said colloidal
platinum are hydrogen and oxygen.

8. The process of claim 6 wherein the electrode medium is platinum coated
on carbon particles and the fuels selected for separate external
saturation of said platinum coated carbon particles are propane and
oxygen.

9. The process of claim 6 wherein the electrode medium for the anode is
nickel boride and the fuel is hydrogen and the electrode medium for
the cathode is colloidal platinum and the oxidant is oxygen.

10. The process of claim 6 wherein the anode medium is colloidal platinum
and the fuel is hydrogen and the cathode is a conventional carbon
block with oxygen as the oxidant.

11. An improved fuel cell with said improvement consisting entirely of having
one or more electrode compartments containing a medium capable of being
reacted to produce electricity, circulated out of the electricity
producing reactive area, reactivated and reintroduced into said electricity
producing reactive area with said electrically conductive medium having
therein the ingredients of a battery such as lead and lead dioxide,
sulfuric acid, and nickel in elemental form, or as their salts, silver,
cadmium, zinc and oxygen, manganese, manganese dioxide and the like being
supplied to the reactive area on the demand for electricity with the
ingredients selected for proper charge and supplied and reacted in polarity
separated but mutually electrically migrative and conductive electrode
compartments together with suitable means to keep the electrolyte volume
and concentration constant and together with suitable means to remove the
electricity so produced.

12. An improved fuel cell electrode compartment having electricity producing
catalyst particles or catalyst coated particles closely packed together
to form a moveable quasi liquid having greater surface area and improved
electrical generating capacity.

13. An improved coin operated and activated time limiting automobile parking
meter and electric outlet for supplying electricity suitable for
recharging automobile fuel cells or storage batteries.

14. An improved coin operated and activated time limiting electric outlet
device for supplying electricity suitable for recharging automobile
fuel cells or storage batteries.

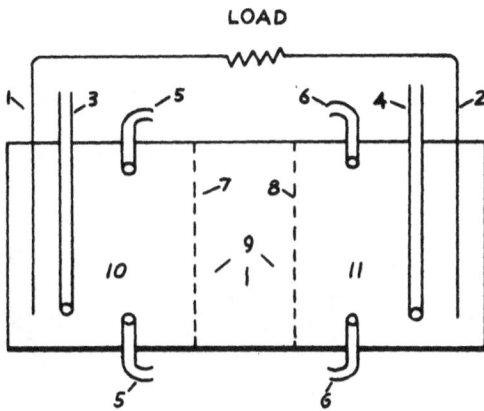

FIG. 1.

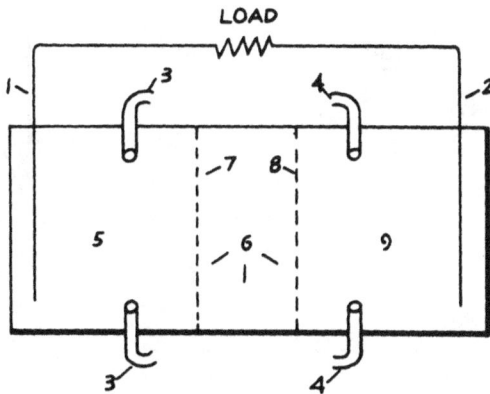

FIG. 2.

INVENTOR.

BY TERRENCE W. PENNER
VERNON L. WAGNER JR.

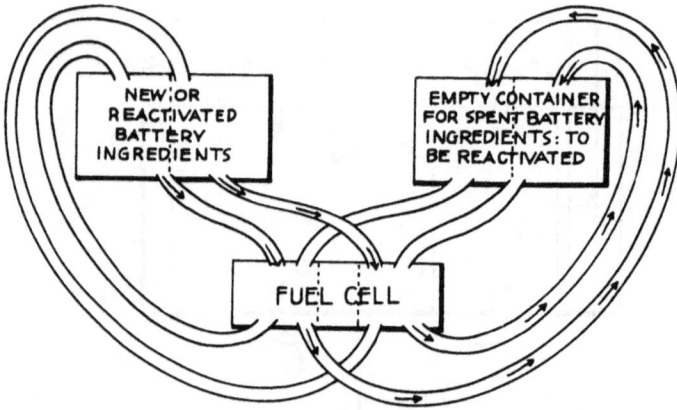

FIG. 3.

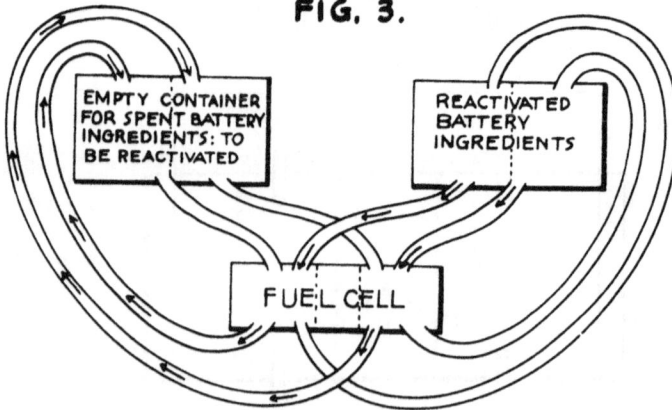

FIG. 3A.

INVENTOR

BY TERRENCE W. FENNER
VERNON L. WAGNER
JR.

FIG. 4.

FIG. 5.

FIG. 6.

INVENTOR.

BY *TERRENCE W. FENNER*
VERNON L. WAGNER
JR.

How Long Does it Take

In general it takes approximately 3 to 6 months from the first letter to the company to establish a contractural relationship until the signing of a contract to buy. The time interval is filled with the submission of the description and disclosure form, the supplying of additional details and the trial testing of the invention by the manufacturer.

Letters, phone calls and possibly a visit also are included in this 6-month interval.

Use the Telephone

The telephone is invaluable in obtaining a friend who may be beneficial to your invention. When a company shows interest, and after several sets of correspondence, try phoning the person you have corresponded with. Here you can explain the workings and answer any questions that may be troubling the person evaluating the invention. Offer your aid in his evaluation.

Try an appointment call. The operator will alert the person that you will call at a certain time (say, in one hour) and place the call for you at that time. This gives the person a chance to get your file and prepare any questions he may have. This call is no more costly than a normal person-to-person call. The system also works well when you want to talk to a Patent Office Examiner after an office action.

The Outright Cash Sale

If the invention is sold outright—for example, an improved part of an existing piece of machinery—the price should reflect the value of the invention to the manufacturer. Naturally, the invention should strive to get the maximum amount possible as this will be the last money he may see for this effort. Let the manufacturer open the bidding to avoid underestimating the value of your invention.

It is recommended, however, that an invention be sold for part cash and part royalty.

How Much Royalty

On a high-volume low cost item having a retail selling price of 10¢ to $1, 2 percent on the price the manufacturer sells the item for is a going royalty rate. Thus, on a $1 item, if the manufacturer received 50¢, the inventor would receive 1¢. This one cent could build up, however, if 5 million units were sold per year   Thus, an inventor could use a sliding scale, of 3 to 4 percent on items retailing for under $10 and up to, say, 25 percent for items costing thousands of dollars—such as a piece of scientific equipment.

Occasionally, an inventor will be paid a unit price per item produced or a smaller royalty in the first year of production because of start up expenses.

Down payments are usually smaller than the inventor would generally like—from $1000 upwards—so it is important to remember the minimum yearly royalty and anti-shelving clauses on any contract you sign to protect your investment. Royalties are usually paid annually or semi-annually. Your contract should stipulate the basis on which royalties are to be paid—yearly, semi-annually, quarterly.

Most manufacturers will want to stipulate who is to be sued if your invention violates an existing patent and who is to sue if another invention infringes upon your patent.

The Value of a Broker

An invention broker is like a real estate agent. He helps you find companies who might be interested in purchasing your invention. He can save you money and possibly earn you more money than you can on your own. Together with your lawyer he can handle the actual sales negotiations. For the inexperienced inventor, the money invested (about $100 plus approximately 10 percent of the sale) is probably well worth the experience and knowledge gained.

The experienced broker over the years has built up contacts with companies, and if he did nothing else than expose your invention to these select contacts, this would be worth the fee.

A good way to check on a broker is to write and ask him to suppy you with a notarized list of the names and addresses of five clients whose inventions he has sold within the last two years.

An invention broker must maintain an office, with secretarial help. compose, copy and merchandise the features of your invention to manufacturers. All this takes time and money, hence the need for brokers fees. The $100 fee isreasonable when all of these things are done and done well.

VIII. How To Manufacture Your Invention

So you are thinking of going into business. To run a business of your own will bring a sense of independence—an opportunity to use your own ideas. You will be top man. You can't be fired. It will mean a chance for higher income because you can collect a salary plus a profit or return on your investment. You will experience a pride in ownership—such as you experience if you own your own home or your own automobile. You can achieve the great satisfaction of building a valuable investment for which there will be a market.

By being top man you can adopt new ideas quickly. Since your enterprise undoubtedly will be a small business—at least in the beginning—you will have no large, unwieldy organization to retrain each time you wish to try something new. If the idea doesn't work you can drop it just as quickly. This opportunity for flexibility will be one of your greatest assets.

These are some of the advantages and pleasures of operating your own business. But let us take a look at the other side.

If you have employees you must meet a payroll week after week. You must always have money to pay creditors—the man who sells you goods or materials, the dealer who furnishes you fixtures and equipment the landlord (if you rent) or the mortgage holder (if you are buying your place of business), the publisher running your advertisements, the tax collector, and many others. You must accept sole responsibility for all final decisions. Wrong judgment on your part can result in losses not only to yourself but, possibly, to your employees, creditors, and customers as well. Moreover, you must withstand, alone, adverse situations caused by circumstances beyond your control, such as depressed economic conditions or strong competition.

The authors wish to thank the U. S. Small Business Administration for permission to edit and publish the information contained in this chapter.

To overcome these disadvantages and keep your business profitable means long hours of hard work. Invariably when you become your own boss you will work longer hours than when you were working for someone else. At least, this will be necessary in the beginning. Then, after all, you will not be entirely your own boss—you must satisfy your customers. Your creditors and your competitors will dictate to you. Health authorities and insurance people will see that you meet certain standards and follow certain regulations. You will have to abide by wage and hour laws and keep records in accordance with the requirements of the tax system.

Are You the Type to Go Into Business

So the first question you should answer after recognizing that there is a dark side as well as a bright side to the prospect of establishing your own business is "Am I the type?"

You will be your most important employee. It is more important that you rate yourself than it is that you rate any prospective employee. From the viewpoint of operating your own business, appraise your strong points and weak points. If you recognize you are weak in salesmanship, for example, you should know it and cover that deficiency by hiring the best sales talent you can afford.

Try to rate yourself objectively. On the next page is a list of 10 traits considered important for the person operating his own business. Add others that you think are significant for the type of business you desire to establish. Then rate yourself. This in no sense is designed as a scientific, psychological test. It is merely for the purpose of calling your attention, a little more vividly than usual, to your own characteristics. After rating yourself, you will do even better by asking a friend to have you rated anonymously by several people who know you. The results may startle you.

Now rate yourself. Be honest. Remember, in starting your own business, you are risking your money and your time.

Are most of your check marks on the left-hand side of the page? That is where they should be. But look them over carefully and be sure none of them is on the left-hand side because of wishful thinking. You will do well to recognize your weak points before opening your business. Perhaps you can compensate for them by hiring the right help or obtaining associates whose strong points offset your weak ones. If you are weak in too many of the traits needed for managing a business,

Rating Scale for Evaluating Personal Traits Important to the Proprietor of a Business

INSTRUCTIONS Place a check mark on the line following each trait where you think it ought to be. The check mark need not be placed directly over one of the guide phrases, because the rating may lie somewhere between the phrases.

Initiative	Additional tasks sought; highly ingenious	Resourceful; alert to opportunities	Regular work peformed without waiting for directions	Routine worker awaiting directions
Attitude toward others	Positive; friendly interest in people	Pleasant, polite	Sometimes difficult to work with	Inclined to be quarrelsome or uncooperative
Leadership	Forceful, inspiring confidence and loyalty	Order giver	Driver	Weak
Responsibility	Responsibility sought and welcomed	Accepted without protest	Unwilling to assume without protest	Avoided whenever possible
Organizing ability	Highly capable of perceiving and arranging fundamentals in logical order	Able organizer	Fairly capable of organizing	Poor organizer
Industry	Industrious; capable of woring hard for long hours	Can work hard but not for too long a period	Fairly industrious	Hard work avoided
Decision	Quick and accurate	Good and careful	Quick, but often unsound	Hesitant and fearful
Sincerity	Courageous, square-shooter	On the level	Fairly sincere	Inclined to lack sincerity
Perseverance	Highly steadfast in purpose; not discouraged by obstacles	Effort steadily maintained	Average determinaion and persistence	Little or no persisence
Physical energy	Highly energetic at all times	Energetic most of the time	Fairly energetic	Below average

do not undertake the venture.

Pick the field you know most about.   The best way to obtain knowledge of a business is through actual experience in it.   If you feel otherwise qualified, but lack sufficient training, seek a job working for somebody else in the business you are considering.   Try to find a position in a well-managed, successful company.   Then absorb as much management know-how as you possibly can.

You will need every minute of experience you can get.   Exactly how much you will need as a minimum depends upon the business and upon your general business knowledge.   Experience in other types of work may teach you something about general business policies and operating methods.   The transfer of experience from one type of business to another is often practicable.

Education will help, too. While there are usually no educational requirements for starting your own business, the more schooling you have had the better equipped you should be.   For example, in most businesses you must know how to figure interest and discounts, keep simple and adequate records, and conduct necessary correspondence. Knowledge of these and many other helpful subjects may be acquired through formal education.

Next, make a sincere effort to determine as best you can whether customers or clients will like the type of business or service you wish to establish.   The business should be in tune with the trend of the times. Choose a field in which expansion is logically expected.   Study surveys and seek advice and counsel.

In conclusion, start with what you are prepared or equipped to offer. What can you do with your present preparation?   Does anyone want the product you can render?   May the product be adapted to present trends in the market?

What are some characteristics associated with success and failure in business?

Managements that combined high levels of education and experience achieved success with striking frequency.   Managements consisting of two or more owners did exceptionally well, particularly if the owners complemented rather than duplicated each other in training and experience.

The following characteristics and practices of management often accompanied success: 5 years or more of managerial experience; consultation before and after starting the business with lawyers, accountants, bankers, future customers, and others; use of such management tools as budgets and controls; and establishment of definite, realistic goals before committing funds.

Failure was *rarely* due to: lack of capital; inability to collect from customers or overextended credit; lack of a market or unsuitability of product; poor records; competition; union trouble or shortage of skilled labor. It was *often* caused by the management's inadequate training, experience, or ability.

In general, the failures were characterized by inability to find and vigorously cultivate an adequate market. Few simple processors or parts manufacturers fail, but few made much profit. On the other hand, manufacturers of products requiring more sophisticated marketing—namely, end products and identifiable components—were more likely to fail, but they also scored most of the outstanding successes.

Opinions about starting a business

Cooperating business founders in a study were asked, "What do you think is the basic difference between people who go into business for themselves and those who always work for someone else?" The major differences suggested lay not in ability or opportunity but in ambition, desire for independence, and willingness to assume responsibilities.

When asked what advice they would offer to people planning to enter business for themselves, a large proportion of the founders recommended obtaining a thorough knowledge of the type of business contemplated. The usual recommendation was for a year or two of experience in the same field before any attempt to start an independent business.

Many cautioned prospective founders to be sure to have adequate capital; and, in the retail field especially, there was strong caution against the extension of credit. Another frequent comment was that a prospective founder must be willing to put in longer hours and work harder than he had as an employee.

Businessmen taking part in the study generally agreed that success in a small business is likely to come to the man who has the following traits:

He works long, hard hours.

He has the ability to recover quickly and press on in the face of a setback.

He is competitive in attitudes and actions.

He is willing to take a minimal profit from his business until he achieves a firm financial position.

He masters the technical and social skills his operation requires.

Additional general leadership characteristics

Factors	Outstanding	Above average	Average	Below average
Ability to work under pressure and changing conditions				
Ability to formulate, present, and obtain acceptance of ideas				
Effectiveness in use of manpower and skills				
Effectiveness of oral expression				
Effectiveness of written expression				
Ability to work harmoniously with subordinates and associates				
Ability to direct and develop employees				
Resourcefulness				
Emotional stability				
Appearance, bearing, and manner				

The enterprising man

The typical entrepreneur differs from the big-business executive in that he cannot live within a framework of occupational behavior set by others.   From his early work experiences, he slowly fashions his conception of how skills, money, equipment, and markets can be brought into a profitable combination.

As the business grows, the founder will cease to be the driving force, and a new generation will take over.   The way of the entrepreneur will no longer be sufficient or necessary.

What motivates the "enterprising man", or entrepreneur, to strike out on his own and set up a new business enterprise?   Is he somehow different from other leadership groups in our society?   Is there a pattern of motives, values, and interests that can be identified as "entrepreneurial"?   The study was a search for answers to these questions.

The background of the entrepreneur

Common patterns of childhood experiences found among 110 entrepreneurs cooperating in the study included (1) the orphaned and alone, (2) the poor but honest, (3) those who came off the farm as opposed to those with urban origins, and (4) those whose fathers had made a tentative step toward entrepreneurship.

Many who had left school at an early age did so because of a feeling of restlessness, of "not getting anywhere." The economic factor, through present, did not have the importance usually attributed to it.

Personality factors

The personality structure of entrepreneurs was examined by a psychologist using the Thematic Apperception Test (TAT). He reported that dominant themes running through the "typical" entrepreneur's personality included the following:(1) a social value system steeped in middle-class mores; (2) lack of social mobility drives (the will to rise in a social hierarchy, achieve power and status); (3) punishing pursuit of tasks and chronic fatigue; (4) lack of problem resolution; (5) satisfactory relations with subordinates on a patriarchal or patronly basis; (6) unwillingness to submit to authority; (7) perception of male authority figures as shadowy, remote beings, not sought out for help or looked up to as models to be emulated.

One point on which the interview material and the TAT analysis agreed closely was the place of adult figures in the world of the entrepreneur. His relation to these figures, more than any other one factor, sets him apart from the men who spend their lives in large organizations and who accept directives handed down by "leaders." The entrepreneur cannot live within a framework of occupational behavior set by others.

Early work experiences

The true school for the entrepreneur, according to the report, is the period between the time when he leaves family life and formal schooling and the time when he firmly establishes himself in his own business. The act of entrepreneurship is that of bringing skills, money, equipment, and markets together into a profitable combination. From his early work experiences, the entrepreneur slowly fashions his conception of how this can be done.

Eventually, according to the report, the typical entrepreneur finds the pattern of his life disrupted. There is a sudden or progressive loss of economic secutity, a loss of the goals and aspirations that have guided him. But it is not the fact of disruption of his life that sets him apart as an entrepreneur. It is that during this time there occurred to him the possibility of establishing a business of his own. In a time of crisis, he does not seek for a situation of security; he goes on into deeper insecurity. In a period of fear and doubt, he finds creativity.

Creation of a new enterprise

The report names four tasks the entrepreneur must accomplish during the creation of a new enterprise.

1. Setting up the firm. In this phase, the entrepreneur is finding the money, men, materials, and equipment necessary for getting operations started.

2. Getting through the knothole. This phase brings long hours of work, low monetary returns, and great uncertainty.

3. Getting rid of partners. In setting up the firm and getting started, the entrepreneur has many times needed financial and other support; and every individual who has furnished support has in one way or another established a foothold within the firm. The entrepreneur must have—and the TAT analysis indicates that he does have—the kind of character the impels him to drive these "intruders" out.

4. On the way at last. During this period, the entrepreneur is engaged in expanding, integrating, and structuring his creation. If he has created imaginatively, organized wisely, and made a sustained effort, his firm will take its place within the network of business enterprise.

About the study

The study is based on interviews with 110 founders of manufacturing enterprises established in Michigan between 1945 and 1958. Forty of them were given the Thematic Apperception Test by a highly trained psychologist. Secondary sources provided information for the comparison of big-business executives and entrepreneurs.

How will you price your products and services	Yes	No
Have you determined what prices you will have to charge to cover your costs and obtain profit?	—	—
Do these prices compare favorably with prices of competitors?	—	—

REFERENCES: *How to Price a New Product: Management Aids Annual No. 3 (45¢ Supt. Docs) MA. 100, Pricing Arithmatic for Small Business Managers; SM 21, Pricing and Profits in Small Stores.*

What selling methods will you use?	Yes	No
Have you studied the sales promotional methods used by competitors?	—	—

Have you outlined your own sales promotion policy?   —   —
Have you studied why customers buy your product
   (service, price, quality, distinctive styling, other)?   —   —
Will you do outside selling?   —   —
Will you advertise in the newspapers?   —   —
Will you do direct mail advertising?   —   —
Will you use posters and handbills?   —   —
Will you use radio and television advertising?   —   —
    REFERENCES: SM 16, *Improving Personal Selling in Small Business;* SM 32, *Methods of Improving Off-Season Sales;* SM 56, *Advertising for Profit and Prestige;* SBB 20, *Advertising-Retail Store*

How will you manage personnel?   Yes   No

Will you be able to hire satisfactory employees,
   locally, to supply skills you lack?   —   —
Do you know what skills are necessary?   —   —
Have you checked the prevailing wage scales?   —   —
Have you a clear-cut idea of what you plan to pay?   —   —
Have you considered hiring someone now employed
   by a competitor?   —   —
Have you checked on the pros and cons of doing so?   —   —
Have you planned your training procedures?   —   —
    REFERENCES: MA 102, *Keeping Your Salesmen Enthusiastic; Sales Training for Small Wholesalers;* in *Marketers Aids Annual No. 1* (45¢ Supt. Docs); SBB 23, *Training Retail Sales People, Sales Training for the Smaller Manufacturer* (20¢ Supt. Docs.)

What records will you keep?   Yes   No

Have you a suitable bookkeeping system ready to
   operate?   —   —
Have you planned a merchandise control system?   —   —
Have you obtained standard operating ratios for your
   type of business to use as guides?   —   —
Have you provided for additional records as necessary?   —   —
Have you a system to use in keeping a check on costs?   —   —
Do you need any special forms?   —   —
Have you made adequate provision for having your
   record keeping done?   —   —

REFERENCES: MA 75, *Protecting Your Records Against Disaster.* in *Management Aids No. 5* (45¢ Supt. Docs.); SM 36, *Picking An Auditor For Your Firm;* SBB 15, *Record Keeping Systems—Small Store and Service Trade.*

Checklist For Going Into Business

People sometimes go into business for themselves without being fully aware of what is involved. Sometimes they are lucky and succeed. More often, they fail because they do not consider one or more of the ingredients needed for business success.

This checklist is designed to help you decide whether you are qualified or have considered the various phases of going into business for yourself. Careful thought now may help you to prevent mistakes and to avoid losing your savings and time later. Use this list as a starter. Consider each question as it have applies to your situation. Check off each question only after you made an effort to answer it honestly. Before you omit a question, satisfy yourself that it does not apply to your particular situation.

After each section, you will find a few references. If you have uncovered doubtful areas or weaknesses in your preparation, it is strongly recommended that you obtain these publications and study them.
You will find it time well spent. Most of the references are available free, on request from any SBA field office or the Small Business Administration, Washington 25, D.C. However, the notation "Supt. Docs." means that the item is for sale at the price indicated by the Superintendent of Documents, Washington 25, D. C. (not from SBA).

Questions to Consider

Are you the type   Yes   No

Have you rated your personal traits such as leadership, organizing ability, perseverance, and physical energy?   —   —
Have you had some friends rate you on them   —   —
Have you considered getting an associate whose strong points will compensate for your weak traits?   —   —
REFERENCES: *Starting and Managing a Small Business of your Own* (40¢ Supt. Docs.), SM 39. *Balanced Skills; Measure of Effective Managers;* SM 46. *Essential Personal*

Qualities for Small Store Managers, SM 52, Are you Really Service-Minded?

What are your chances for success	*Yes*	*No*
Have you had any actual business experience?	—	—
Have you obtained some basic management experience working for someone else?	—	—
Have you analyzed the recent trend of business conditions (good or bad?)	—	—
Have you analyzed business conditions in the city and neighborhood where you want to locate?	—	—
Have you analyzed conditions in the line of business you are planning?	—	—

Choosing the Legal Structure for Your Firm

If a small concern is to operate effectively in today's climate and continue to exist when the present owners can no longer function, its legal structure must be right.

Joe Porter's problem

Take, for instance, the case of Joe Porter. In 1945, Joe took his wartime savings, a chunk of his wife's nest egg, and a loan from his bank and started a small plastics molding business. He learned fast, worked long hours, and got some good breaks. Today, he owns an up-to-date little factory with good equipment, 97 employees, and annual sales of $940,000.

Until lately, Joe had not been interested in making any changes in the legal structure of his firm. He enjoyed being sole owner. He liked being able to make independent decisions and felt that there was really no need to risk "upsetting the apple cart" by revamping his organization.

Nevertheless, Joe listened carefully when his accountant brought up the effects of sticking to the status quo. For one thing, tax considerations backed up the point. For 1964, Joe's firm made a net profit of $47,000. His personal and family expense had totalled $14,000. All the same, he had to pay personal income tax on the entire business profit. He reported his income on the calendar year basis and filed a joint income tax return with his wife. After subtracting $4,000 for exemptions and deductions, he reported $43,000 on which he paid

$14,415 tax. However, if the firm had been a corporation which reported its income on the calendar year basis and he had taken, say, a $17,000 salary, his tax bill would have looked like this:

1.	Personal income tax on $17,000 salary (less $4,000 exemptions and deductions)	$ 2,690

2.	Corporate taxes on the net profit
	(after salary) of $30,000:
	    22% on $30,000
	    28% on 5,000
		
		6,600
		1,400

3.	Total tax	$10,690

Thus, if the company had been incorporated, Joe would have saved $3,725 in taxes for last year alone. That, said the accountant, seemed worthwhile. Moreover, there would have been no objections from the Internal Revenue Service to a $17,000 salary for the president of a business the size of Joe's. Many comparable executives get as much or more for services they actually performed.

With these facts in mind, Joe asked his lawyer for a fill-in on legal structure in general so that he could better decide whether he should go further in making changes in his company's set-up.

Three main choices

Broadly speaking, the lawyer said, there are three principal kinds of business:

Proprietorship—which is the easiest to begin and end (sometimes prematurely), can have the most flexible purpose for its operations, needs no Government approval, has business profits taxed as personal income, and makes the owner personally liable for debts and taxes.

Partnership—which is the simplest for two or more people to start and terminate, has the same flexibility of objective, has partners taxed separately, and makes personally liable for debts and taxes all except limited partners.

Corporation—which is the most formal of structures, operates under State laws, has continuous and separate legal life, has its scope of activity and name restricted by a charter, has the business profits taxed separately from earnings of executives and owners, and makes only the company (not the owners nor managers) liable for its debts and taxes.

(There are other types of legal structure such as syndicates, joint stock companies, Massachusetts trusts, and pools, the lawyer pointed out. However, these are specialized and rare. For that reason they are eliminated from this discussion.)

Six points to check

In analyzing your own situation, it pays to go to the expense of getting advice and guidance from competent legal counsel. Great care should be taken to make the right decision the first time. Among other things the lawyer pointed out, it is worthwhile for the top executive to be familiar with the highlights of six main points on legal structure in addition to tax considerations: (1) Costs and procedures in starting; (2) size of risk—that is, amount of investors' liability for debts and taxes; (3) continuity of the concern; (4) adaptability of administration; (5) influences of applicable laws; and (6) attraction of additional capital.

1. *Costs and procedures in starting*

Single proprietorships are the easiest to get started. The costs of formation are low. Basically, all you have to do is to find out whether you need a license to carry on your particular business and whether you have to pay a State tax or license fee.

General partnerships are also started quite simply. You can set one up by having the executives in the business sign what is called a partnership agreement. A written document, however, is not necessarily a prerequisite, since an oral agreement can be equally effective. Moreover, a partnership may even be implied by actions which the managers of an unincorporated business have taken—even though no agreement of any kind, oral or written, exists.

Limited partnerships are somewhat more difficult to set up. To form one you file with the proper State official a written contract drawn according to certain legal requirements. This contract permits you to limit the liability of one of more of the partners to just the amount they invested. But you must designate at least one general partner in addition to the limited partners. And all limited partners must have actually invested in the partnership. According to the Uniform Limited Partnership Act, those investments may be either cash or tangible property, but not services. Lastly, you must conform strictly to the laws of the particular state in which you organize; otherwise your business will be considered as a general partnership.

Corporations are more complicated to form than any of the other types of organization. You can create one only by following strictly the legal procedures of the particular state in which the corporation is being set up. First, certain responsible people are needed to organize and become officials in the new corporation. Next, they must file with the designated state official a special document called the "articles of incorporation." Then they must pay an initial tax and certain fiiling fees. And finally, in order to do the business for which the corporation was formed, various official meetings must be conducted to deal with specified details of organization and operation.

2. The size of the risk

The degree to which investors in your enterprise risk legal liability for the debts of the business is a cardinal consideration. Regardless of legal structure, creditors are always entitled to be paid out of business assets before any equity capital may be withdrawn. In cases where those assets are insufficient, the extent to which owners can be compelled to meet creditors' claims out of their own pockets varies with the type of organization.

A single proprietor is personally liable for all debts of his business —to the extent of his entire property. He cannot restrict his liability in any way. Likewise, each member of a *general partnership* is himself fully responsible for all debts owed by his partnership—irrespective of the amount of his investment in the business. In the *limited partnership*, however, the limited partners are protected; they risk only the loss of the capital they have invested. But the general partners in a limited partnership are liable jointly and severally for all debts just like any other general partner. And remember, there must be at least one general partner in any limited partnership.

Corporations have a real advantage, as far as risk goes, over other legal structures. Creditors can force payment on their claims only to the limit of the company's assets. Thus, while a shareholder may lose the money he put into the company, he cannot be forced to contribute additional funds out of his own pocket to meet business debts. This is true even though the corporate assets may be insufficient to meet creditors' claims.

3. Continuity of the concern

In choosing the legal structure for your business, you should also understand clearly how it influences the continuity of the business. Although *single proprietorships* have no time limit on them by law, they are not fundamentally perpetual. Illness of the owner may derange the business and his death ends it. *Partnerships* are perishable in the same general sense—since they are terminated by the death or withdrawl of any one of the partners.

Corporations have the most permanent legal structure of all. They have a separate continuous life of their own. The withdrawl, insolvency, injury, illness, or death of a person officially concerned in a corporation does not mean its finish. Moreover, the certificates of stock, which represent investments and ownership in the business, may be transferred from one person to another without hampering the concern's operations.

4.   *Adaptability of administration*

In the *single proprietorship*, policy and operations rest, of course, in one individual. This situation can be both good and bad. On the one hand, concentration of managements avoids the problems of opposing factions and divided responsibilities. The fact that the chief executive is in full charge, and is in complete control of profits, can be an incentive to careful management. On the other hand, many a man is not competent to handle all management jobs himself. To be sure, an owner can, and often does, employ assistants to whom he assigns various details. But he still reaps the rewards or the penalties of what they do. It is also worth noting that after incorporating, the owner of a small business does not necessarily lose control of the enterprise. In many small, closely held corporations, the former sole owner can and often does retain control by the ownership of a majority of the stock in the newly formed corporation.

In *general partnerships*, each partner typically has an equal role in administration, with the various operating functions divided among them. The combined abilities and knowledge of several executives gives the partnership an advantage over the single proprietorship. But the division of functional responsibility among the several partners may lead to fundamental policy disagreements. Compared with corporations, partnerships have the following administrative features: Decisions may be taken and changes adopted simply by oral agreement among the partners. In limited partnerships, the limited partners may not engage in management functions; if they do, they may be held fully liable as general partners. They are, however, entitled to inspect the books and obtain full and complete information regarding the business.

In *corporations*, the stockholders do not necessarily participate either in operations or in policy formulation, but they may. Often, however, those functions are centralized in a relatively small group of executives who own only a small percentage of the shares. Although corporations can get away from the shortcomings of the limited ability or knowledge of one person, they do run some risk of inefficient management where those in control have little or no direct financial interest. Corporations have an advantage over partnerships in this way: In partnerships each partner can act as general agent for the business; but in corporations, the stockholders cannot bind the firm by their acts just because they have invested capital in it.

5. *Influences of applicable laws*

Single proprietorship is the oldest and most widespread legal structure of business. As a result, little doubt remains as to the influences of laws regulating its legal rights, and obligations. Likewise the relationships are clear between a sole owner, his agents, his creditors, and others with whom he deals in business. A private citizen working in Iowa, can carry on business in Kansas without paying any greater taxes or incurring any more obligations in Kansas than local Kansas businessmen have.

Broadly speaking, this same situation is also true for a *partnership*. Of course, a state may require the purchase of a license to carry on a particular kind of business. But the license will be equally available to businessmen of any state so long as they conform to prescribed uniform standards. (This equality of opportunity derives from the United States Constitution which guarantees to citizens of each state "all privileges and immunities" provided to citizens of the other states.) Thus, the legal structures which do not involve any artificial entity (as a corporation does) provide a freedom of action in all states that corporations cannot match.

Corporations owe their legal life solely to the states in which they are organized. No other state is required to recognize them. To be sure, all states do permit out-of-state corporations to function inside their boundaries. Nevertheless, out-of-state corporations must always comply with special in-state obligations such as (1) filing certain legal papers with the proper state officials; (2) appointment of a representative in the state to act as agent in serving process on the "foreign" corporation; and (3) payment of specified fees and taxes.

Also, corporations are regulated by numerous state laws which vary considerably. Even when the language is similar, these laws can be, and have been, interpreted differently in different places. Therefore, in running a corporation effectively, competent legal counsel is virtually indispensable. The normal course of business, for example, can easily involve statutes and court decisions of a state other than the one where the corporation was founded. Nevertheless, the essential feature of limited liability of stockholders is preserved in every state.

6. *Attraction of Additional Capital*

Every business may require additional funds from time to time to carry on operations. And if it can not obtain adequate capital, it may well be headed for failure. It is important, therefore, in deciding upon legal structure to take into account the means for attracting new money.

In *single proprietorship*, the owner may raise additional money by borrowing, by purchasing on credit, and by investing additional amounts himself. Since he is personally liable for all the debts of his business, banks and suppliers will look carefully at his personal wealth. Consequently, the funds he can get will always be limited by his own circumstances. For this reason alone, a business requiring large amounts of capital for successful operation should probably not be organized as a single proprietorship.

Partnerships can often raise funds with greater ease, since the resources of all partners are combined in a single undertaking. Like single proprietors, partners must accept full personal liability for business debts; for this reason, a partnership may be able to borrow on better terms than some corporations. In addition, outsiders may be willing to extend credit because of the security deriving from the individual partners' full liability.

Corporations are usually in the best position of all to attract capital. They may, for example, acquire additional funds by borrowing money by pledging corporate assets. Also, they may sell securities to the public and attract a wide range of investors. A shareholder's investment in a corporation will not subject him to any financial risk beyond the amount of his holdings. In addition, as a part owner, he has the prospect of sharing directly, through dividends and rising value of the securities, in any profits the concern makes.

Tax effects of the form of organization of a multiple-ownership firm may differ for individual owners of the firm. This difference, if recognized, can be compensated for by salary allowances or other means.

The Subchapter S tax-option plan has resolved many tax inequities, but expert knowledge of its application is essential. Upon making the change to another form of business a taxable transfer may bring net tax savings in the long run. If working capital is short, however, a nontaxable transfer may be preferable.

Congress has given small-business owners a number of alternatives intended to prevent Federal income taxes from imposing undue burdens on small firms. One of these choices is whether the business shall be a proprietorship, a partnership, a conventional corporation, or a corporation that elects the tax plan allowed by Subchapter S of the Internal Revenue Code.

Subchapter S tax option

Many firms could have obtained the lower tax of the partnership form while retaining the business advantages of the corporate form by taking advantage of the Subchapter S taxoption. Subchapter S of the Internal Revenue Code, enacted in 1958, allows corporations to elect to pay taxes in a manner similar to partnerships if they have no more than 10 stockholders and meet other qualifications.

The tax-option provision has resolved many tax inequities, according to the report, and it fits the operational plan of the small business well. But it has many pitfalls. The businessman who chooses this form must be in constant touch with both an accountant and an attorney who have made a careful study of this section of the tax law.

Losses

Businesses that have widely fluctuating earnings pay more tax over a period of years than comparable firms earning at a more uniform rate. The net-operating-loss carryback and carryover provisions of the Internal Revenue Code lessen this difference somewhat. They allow an operating loss to be "carried back" for 3 years for refunds of taxes paid in those years. If the loss is not used up by the carrybacks to these 3 years, the rest can be carried forward for a maximum of 5 years. Operating losses can thus be spread over 9 years.

Conflicting Goals

Individual owners of a multiple-ownership firm may have widely differing amounts and types of outside income. As a result, the form of organization or method of transferring property to the new organization that is most advantageous for one owner may be undesirable from another owner's point of view. However, if the effects of the various alternatives are projected before a choice is made, the inequity can be compensated for by salary allowances or other means.

Should you share ownership with others?   Yes   No

Do you lack needed technical or management skills
   which can be most satisfactorily supplied by one or
   more partners?   —   —

Do you need the financial assistance of one or more
   partners?   —   —

Have you checked the features of each form or
   organization (individual proprietorship, partnership,
   corporation) to see which will best fit your situation?   —   —

REFERENCES: MA80, *Choosing the Legal Structure for Your Firm;*
MA 111, *Steps in Incorporating a Business, Equity Capital and Small Business* (35¢ Supt. Docs.)

Say that you have now decided that you are the type who can operate a business of your own. You have given some attention to the overall chances for success and have chosen the business you wish to establish. What are some of the practical problems of starting the business? How much capital will you need? Where can you obtain it?

How much capital?

First, how much capital will you need? Your answer to this question deserves careful study and investigation. No average figure can be specified, since the amount differs widely; depending on the kind of business, type of establishment, location, current price level, and other factors.

The procedure for estimating capital requirements for a small factory, is as follows: For example, suppose you wish to operate a factory to manufacture a golf practice game and to make $10,000 net profit annually. A $280,000 sales volume is necessary (figured at 3.6 percent net profit before taxes). How many units must be produced to attain this volume? Suppose you plan to manufacture the games which you will sell at an average price of $10. To obtain $280,000 sales volume you must sell 28,000 units. This means an average of 560 units per week for 50 weeks. How many machines will be required to produce 560 per week? How much other equipment? How much down payment for the equipment will be necessary? How many operators will be needed? Answers to these and other questions can be obtained from equipment suppliers.   You must add to the down payment for equipment estimates of costs for materials, wages, rent, sales, office and other expenses for a period necessary to produce enough units for one complete turn; that is, the annual production (28,000 units in this case) divided by the expected number of stock turns per year.   Forms and suggestions to help in estimating these costs may be found in Chapter 12, "Pitfalls in Estimating Your Manufacturing Costs," *Management Aids for Small Business: Annual No. 2.*

Estimating capital requirements for a service establishment will involve a combination of the methods used for merchandising and manufacturing businesses.   To the extent that the service business carries goods for resale, capital requirement estimates could be made in the manner outlined for wholesaling and retailing concerns.   To the extent that it sells labor or machine work, capital for equipment and wages could be estimated in much the same way as for a factory.

Your available capital must exceed the initial capital requirements (as computed on your work sheet) by a safe margin.   It is necessary not only to have money to get started but also enough in reserve to carry the business until it becomes self-supporting.   In some instances this time may be from 4 to 6 months; in others it may be even more.   If you do not have sufficient capital, remember you may be unable to:

1.   Afford enough employees to keep the business operating.
2.   Invest in proper equipment.
3.   Maintain an adequate stock of merchandise or materials in order to build sales volume.
4.   Take advantage of discounts offered by creditors and, thereby, be burdened with heavy interest penalties.
5.   Grant customer credit to meet competition.

Getting the money

Now that you have computed your initial capital requirements, where will you get the money?   The first source is your personal savings.   Then relatives, friends, or other individuals may be found who are willing to "venture" their savings in your business.   Before obtaining too large a share of the capital from outside sources, remember that you should have personal control of enough to assure yourself ownership.   Once you can show that you have carefully worked out your financial requirements, and can demonstrate experience and integrity, a lending institution may be willing to finance part of your operating

capital needs.   This may be done on a short term basis of 60 days to
as much as 1 year.

The main outside sources of capital credit in the early days of your
business are (1) the commercial bank and (2) the trade creditor or
equipment manufacturer.   Other sources, such as small loan companies,
factoring companies, commercial credit companies, sales finance companies, and insurance companies will not be discussed here.   Loan
sources in the Federal Government are discussed in Chapter 4, of the
Small Business Administration publication *Management Aids for Small Business: Annual No. 2.*

It is well to get acquainted with a banker.   Even if you do not need
his assistance at the outset, he may be helpful at some later time.
"How to Choose Your Banker Wisely," Chapter 10 of *Management Aids
for Small Business: Annual No. 2* give helpful advice about this.

When you deal with your banker, sell yourself.   Openly discuss your
plans and difficulties with him.   It is his business not to betray confidence.   If you need financial assistance, carefully prepare, in written
form, complete information so that anyone approached for aid may gain
a thorough understanding of your entire proposition.   Many businessmen or prospective business operators have destroyed their chances
of obtaining financial help by the failure to present their proposition
properly.   For helpful information on this subject, consult Chapter
11, "Borrowing Money from Your Bank," *Management Aids for Small
Business: Annual No. 2.*

The companies from which you buy equipment or merchandise may
also furnish you capital in the form of credit.   Manufacturers of equipment, such as store fixtures, cash registers, and industrial machinery,
frequently have financing plans under which you may buy on the installment basis and subsequently pay for the equipment out of income.
Moreover, the wholesalers or suppliers from which you purchase
merchandise extend credit.   You are not required to pay for the goods
at once.   If goods are for resale, no security other than repossession
rights of the unsold goods is involved.   However, too extended use
use of such credit may prove expensive.   Usually cash discounts are
quoted if a bill is paid within 10, 30, or 60 days.   For example, a term
of sale quoted as "2-10; net 30 days," means that a cash discount of 2
percent will be granted if the bill is paid within 10 days.   If not paid in
10 days, the entire amount is due in 30 days.   If you do not take advantage of the cash discount, you are paying 2 percent to use money for
20 days, or 36 percent per year.   This is high interest.

One of the principal causes of failures among businesses is inadequate financing. If you do go into business, remember it is your responsibility to provide, or obtain from others, sufficient capital to supply a firm foundation for the enterprise. An excellent booklet on this subject is *A Handbook of Small Business Finance*, Small Business Management Series, S.B.A.

Communities under "Community Development Laws" often will lend money or build plants to entice manufacturers to locate in their town. This source of capital should be thoroughly evaluated.

The ABC of Borrowing

Some small businessmen cannot understand why a lending institution refuses to lend them money. Others have no trouble getting funds, but they are surprised to find strings attached to their loans. Such owner-managers fail to realize that banks and other lenders have to operate by certain principles just as do other types of business. Inexperience with borrowing procedures often creates resentment and bitterness. The stories of three small businessmen illustrate this point.

"I'll never trade here again," Bill Smith said when his bank refused to grant him a loan. "I'd like to let you have it, Bill," the banker said, "but your firm isn't earning enough to meet your current obligations." Mr. Smith was unaware of a vital financial fact, namely, that lending institutions have to be certain that the borrower's business can repay the loan.

Tom Jones lost his temper when the bank refused him a loan because he did not know what kind or how much money he needed. "We hesitate to lend," the banker said, "to businessmen with such vague ideas of what and how much they need."

John Williams' case was somewhat different. He did not explode until after he got the loan. When the papers were read to sign, he realized that the loan agreement put certain limitations on his business activities. "You can't dictate to me," he said and walked out of the bank. What he did not realize was that the limitations were for his good as well as for the bank's protection.

Knowledge of the financial facts of business life could have saved all three men the embarrassment of losing their tempers. Even more important, such information would have helped them to borrow money at a time when their businesses needed it badly.

Is your firm credit worthy?

The ability to obtain money when you need it is as necessary to the operation of your business as is the right equipment, reliable sources of supplies and materials, or an adequate labor force.   Before a bank or any other lending agency will lend you money, the loaning officer must feel satisfied with the answers to the five following questions:

1.   What sort of person are you, the prospective borrower?   By all odds, the character of the borrower comes first.   Next is his ability to manage his business.

2.   What are you going to do with the money?   The answer to this question will determine the type of loan—short-or long-term.   Money to be used for the purchase of seasonal inventory will require quicker repayment than money used to buy fixed assets.

3.   When and how do you plan to pay it back?   Your banker's judgment as to your business ability and the type of loan will be a deciding factor in the answer to this question.

4.   Is the cushion in the loan large enough?   In other words, does the amount requested make suitable allowance for unexpected developments?   The banker decides this question on the basis of your financial statement which sets forth the condition of your business and/or on the collateral pledge.

5.   What is the outlook for business in general and for your business particularly?

Adequate financial data is a "must"

The banker wants to make loans to businesses that are solvent, profitable, and growing.   The two basic financial statements he uses to determine those conditions are the balance sheet and profit-and-loss statement.   The former is the major yardstick for solvency and the latter for profits.   A continuous series of these two statements over a period of time is the principal device for measuring financial stability and growth potential.

In interviewing loan applicants and in studying their records, the banker is especially interested in the following facts and figures.

General information

Are the books and records up to date and in good condition?   What is the condition of accounts payable?   Of notes payable?   What are the salaries of the owner-manager and other company officers?   Are all taxes being paid currently?   What is the order backlog?   What is the number of employees?   What is the insurance coverage?

Accounts receivable

Are there indications that some of the accounts receivable have already been pledged to another creditor?   What is the accounts receivable turnover?   Is the accounts receivable total weakened because many customers are far behind in their payments?   Has a large enough reserve been set up to cover doubtful accounts?   How much do the largest accounts owe and what percentage of your total accounts does this amount represent?

Inventories

Is merchandise in good shape or will it have to be marked down? How much raw material is on hand?   How much work is in process? How much of the inventory is finished goods?   Is there any obsolete inventory?   Has an excessive amount of inventory been consigned to customers?   Is inventory turnover in line with the turnover for other businesses in the same industry?   Or is money being tied up too long in inventory?

Fixed assets

What is the type, age, and condition of the equipment?   What are the depreciation policies?   What are the details of mortgages or conditional sales contracts?   What are the future acquisition plans?

What Kind of Money

When you set out to borrow money for your firm, it is important to know the kind of money you need from a bank or other lending institution.   There are three kinds of money: short-term money, term money, and equity capital.

Keep in mind that the purpose for which the funds are to be used is an important factor in deciding the kind of money needed.   But even so, deciding what kind of money to use is not always easy.   It is sometimes complicated by the fact that you may be using some of various kinds of money at the same time and for identical purposes.

Keep in mind that a very important distinction between the types of money is the source of repayment.   Generally, short-term loans are repaid from the liquidation of current assets which they have financed.   Long-term loans are usually repaid from earnings.

Short term bank loans

You can use short-term bank loans for purposes such as financing accounts receivable for, say, 30 to 60 days. Or you can use them for purposes that take longer to pay off—such as for building a seasonal inventory over a period of 5 to 6 months. Usually, lenders expect short-term loans to be repaid after their purposes have been served: for example, accounts receivable loans, when the outstanding accounts have been paid by the borrower's customers, and inventory loans, when the inventory has been converted into saleable merchandise.

Banks grant such money either on your general credit reputation with an unsecured loan or on a secured loan—against collateral.

The *unsecured loan* is the most frequently used form of bank credit for short-term purposes. You do not have to put up collateral because the bank relies on your credit reputation.

The *secured loan* involves a pledge of some or all of your assets. The bank requires security as a protection for its depositors against the risks that are involved even in business situations where the chances of success are good.

Term borrowing

Term borrowing provides money you plan to pay back over a fairly long time. It is broken down into two forms: (1) intermediate— loans longer than 1 year but less than 5 years, and (2) long-term—loans for more than 5 years. However, for your purpose of matching the kind of money to the needs of your company, think of term borrowing as a kind of money that you probably will pay back in periodic install- ments from earnings.

Equity capital

Some people confuse term borrowing and equity (or investment) capital. Yet there is a big difference. You do not have to repay equity money. It is money you get by selling a part interest in your business. You take people into your company who are willing to risk their money in it. They are interested in potential income rather than in an im- mediate return on their investment.

How Much Money

The amount of money you need to borrow depends on the purpose for which you need funds. Figuring the amount of money required for business construction, conversion, or expansion—term loans or equity capital—is relatively easy. Equipment manufacturers, architects, and builders will readily supply you with cost estimates. On the other hand, the amount of working capital you need depends upon the type of business you are in. While rule-of-thumb ratios may be helpful as a starting point, a detailed projection of sources and uses of funds over some future period of time—usually for 12 months—is a better approach. In this way, the characteristics of the particular situation can be taken into account. Such a projection is developed through the combination of a predicted budget and a cash forecast.

The budget is based on recent operating experience plus your best judgment of performance during the coming period. The cash forecast is your estimates of cash receipts and disbursements during the budget period. Thus, the budget and the cash forecast together represent your plan for meeting your working capital requirements. To plan your working capital reqjirements, it is important to know the "cash flow" that your business will generate. This involves simply a consideration of all elements of cash receipts and disbursements at the time they occur.

Profit and Loss Statement Adapted to Show Cash Flow

The profit and loss statement elements listed below have been adapted to show cash flow.   Note that it shows "Bank Loans To Be Obtained" as well as "Bank Loans To Be Repaid."   The P and L statement should be projected for each month of the year.

Monthly Operations

Net Sales	$_____
Less: Material used	
Direct labor	$_____
Other manufacturing expense	$_____
Cost of goods sold	$_____
Gross profit	$_____
Less: Sales expense	$_____
General and administrative expense	$_____
Operating profit	$_____

Cash Flow

Cash balance (beginning)	$_____
Receipts from receivables	$_____
Total available cash	$_____
Less Disbursements	
Trade payables	$_____
Direct labor	$_____
Other manufacturing expense	$_____
Sales expense	$_____
General and administrative expense	$_____
Fixed asset additions	$_____
Bank loans to be repaid	$_____
Total disbursements	$_____
Indicated cash shortage	$_____
Bank loans to be obtained	$_____
Cash balance (ending)	$_____
Materials purchased	$_____
Month-end position	
Accounts receivable	$_____
Inventory	$_____
Accounts payable	$_____
Bank loans payable	$_____

What Kind of Collateral

Sometimes your signature is the only security the bank needs when making a loan. At other times, the bank requires additional assurance that the money will be repaid. The kind and amount of security depends on the bank and on the borrower's situation.

If the loan required cannot be justified by the borrower's financial statements alone, a pledge of security may bridge the gap. The types of security are: endorsers, co-makers, and guarantors; assignment of leases; trust receipts and floor planning; chattel mortgages; real estate; accounts receivables; savings accounts; life insurance policies; and stocks and bonds. In a substantial number of states where the Uniform Commercial Code has been enacted, paperwork for recording loan transactions will be greatly simplified.

Endorsers, co-makers, and guarantors

Borrowers often get other people to sign a note in order to bolster their own credit. These *endorsers* are contingently liable for the note they sign. If the borrower fails to pay up, the bank expects the endorser to make the note good. Sometimes, the endorser may be asked to pledge assets or securities that he owns.

A *co-maker* is one who creates an obligation jointly with the borrower. In such cases, the bank can collect directly from either the maker or the co-maker.

A *guarantor* is one who guarantees the payment of a note by signing a guaranty commitment. Sometimes, a manufacturer will act as guarantor for one of his customers.

Assignment of leases

The assigned lease as security is similar to the guarantee. It is used, for example, in some franchise situations. The bank lends the money on a building and takes a mortgage. Then the lease, which the dealer and the parent franchise company work out, is assigned so that the bank automatically receives the rent payments. In this manner, the bank is guaranteed repayment of the loan.

Warehouse receipts

Banks also take commodities as security by lending money on a warehouse receipt. Such a receipt is usually delivered directly to the bank and shows that the merchandise used as security either has been placed in a public warehouse or has been left on your premises under the control of one of your employees who is bonded (as in field warehousing). Such loans are generally made on staple or standard merchandise which can be readily marketed. The typical warehouse receipt loan is for a percentage of the estimated value of the goods used as security.

Trust receipts and floor planning

Merchandise, such as automobiles, appliances, and boats, has to be displayed to be sold. The only way many small marketers can afford such displays is by borrowing money. Such loans are often secured by a note and a trust receipt. This trust receipt is the legal paper for floor planning. It is used for serial-numbered merchandise. When you sign one you (1) acknowledge receipt of the merchandise, (2) agree to keep the merchandise in trust for the bank, and (3) promise to pay the bank as you sell the goods.

Chattel mortgages

If you buy equipment such as a cash register or a delivery truck, you may want to get a chattel mortgage loan. You give the bank a lien on the equipment you are buying. The bank also evaluates the present and future market value of the equipment being used to secure the loan. How rapidly will it depreciate? Does the borrower have the necessary fire, theft, property damage, and public liability insurance on the equipment? The banker has to be sure that the borrower protects the equipment.

Real estate

Real estate is another form of collateral for long-term loans. When taking a real estate mortgage, the bank finds out (1) the location of the real estate, (2) its physical condition, (3) its foreclosure value, and (4) the amount of insurance carried on the property.

Accounts receivable

Many banks lend money on accounts receivable. In effect, you are counting on your customers to pay your note.

The bank may take accounts receivable on a *notification* or a *nonnotification* plan. Under the *notification* plan, the purchaser of the goods is informed by the bank that his account has been assigned to it and he is asked to pay the bank. Under the nonnotification plan, the borrower's customers continue to pay him the sums due on their accounts and he pays the bank.

Savings accounts

Sometimes, you might get a loan by assigning to the bank a savings account.   In such cases, the bank gets an assignment from you and keeps your passbook.   If you assign an account in another bank as collateral, the lending bank asks the other bank to mark its records to show that the account is held as collateral.

Life insurance

Another kind of collateral is life insurance.   Banks will lend up to the cash value of a life insurance policy.   You have to assign the policy to the bank.   If the policy is on the life of an executive of a small corporation, corporate resolutions must be made authorizing the assignment.   Most insurance companies allow you to sign the policy back to the original beneficiary when the assignment to the bank ends.

Some people like to use life insurance as collateral rather than borrow directly from insurance companies.   One reason is that a bank loan is often more convenient to obtain and usually may be obtained at a lower interest rate.

An SBA Loan

The business loan program of the Small Business Administration is designed to provide needed financing to creditworthy small businesses when loans are not available to them on reasonable terms from private lending sources.   The primary purpose of this financial assistance is to provide small firms with funds to purchase equipment and materials, to expand and modernize operations, or to use as working capital.

The SBA business loan program is not competitive with banks or other private lending institutions.   The program is based on cooperation with private lending institutions in meeting the credit needs of small business, and many banks participate with SBA in loans to worthy small concerns.   This cooperative plan not only assists with the immediate financial needs of small firms, but also helps them establish credit with banks in their own communities.

The SBA's loans are of two types—*participation* and *direct*.   In a participation loan, the Agency joins with a bank (or other private lending institution) in a loan to a small business concern.   In a direct loan, there is no participation by a private lender—the loan is made entirely and directly by SBA to the borrower.   By law, the Agency may not make a direct loan if a participation loan can be arranged.

The Agency's participation may be on either a *deferred* or an *immediate* basis.   When it participates on a deferred basis, SBA agrees to purchase from the bank, at any time during a stated period, a fixed percentage of the outstanding balance of the loan.   When it participates in a loan on an immediate basis, it purchases immediately from the bank a fixed percentage of the original principal balance of the loan. However, under the Small Business Act, the Agency may not enter into an immediate participation if a deferred participation is available.

Regular Participation and Direct Loans

The maximum Small Business Administration share of a Regular Participation Loan, on a percentage basis, is 90 percent; on a dollar basis, $350,000.   Application for a Regular Participation loan is made on SBA Form 4, available from any Agency field office.   To apply for a Regular Participation loan, the small business should file with the bank that proposes to participate in the loan three copies of the application and three copies of any supporting documents.   The bank then should prepare a request to SBA for a participation agreement using for this purpose space provided on the SBA form.   The small business or the bank should then file two copies of the application and supporting documents with the SBA office serving the territory in which the applicant's home office is located.   The third copy should be retained by the bank.

The maximum SBA direct loan to any one small business also is $350,000.   To apply for a direct loan, a small business should file two copies of its application (using SBA Form 4) and supporting documents with the SBA office serving the area.   The application must be accompanied by a letter from the small concern's bank stating it is unable to make or participate in the loan.   Where the bank has declined the loan because the requested amount would exceed its legal lending limit or is greater than the bank normally lends to any one borrower, the applicant must be able to show also that the loan is not obtainable from a correspondent bank of his bank of account or from another lending institution whose lending capacity is adequate to cover the requested loan.   In any event, if the small business is located in a city with a population of 200,000 or more, its application must be accompanied by letters from two banks stating that they cannot make or participate in the requested loan.

Long Terms

Small Business Administration business loans generally are repayable in regular monthly installments, including interest on the unpaid balance. Interest is charged only on the actual amount borrowed, and for the actual time the money is outstanding.   All or any part of a loan may be repaid without penalty before it is due.

The maximum maturity of a loan generally is 10 years.   However, loans for working capital usually are limited to 6 years; loans for construction purposes may have a maturity of 10 years plus the estimated time required to complete construction, and loans made to small business pools for construction of facilities may have a maturity of up to 20 years.

The interest rate on SBA's direct business loans, and the maximum interest rate on the Agency's share of a participation loan is 5 1/2 percent.   A private lending institution may set a higher rate than 5 1/2 percent on its share of a participation loan, provided the rate is legal and reasonable.

An SBIC Loan

An SBIC is a privately owned and privately operated small business investment company licensed by the Small Business Administration to provide equity capital and long-term loans to small firms.   Often, an SBIC also provides management assistance to the companies it finances.

Small businesses generally have difficulty obtaining long-term capital to finance their growth.   Prior to 1958, there were few places a small company could turn for money once it had exhausted its secured line of credit from banks or SBA.   To help close this financing gap, Congress passed the Small Business Investment Act of 1958 which authorized SBA to license, regulate and help finance privately organized and privately operated SBIC's.

Today there are more than 700 of these companies located in all parts of the country.   Their individual capital ranges from $300,000 to about $25 million.   As an industry, the SBIC's have total assets of about $750 million, and additional funds are available to them through borrowings from SBA and private sources. By the end of 1964, the companies had invested more than $500 million in small businesses.

Most SBIC's are owned by relatively small groups of local investors. However, the stock of nearly 50 SBIC's is publicly traded, more than 80 SBIC's are partially or wholly owned by commercial banks, and some SBIC's are subsidiaries of other corporations.

SBIC financing and its cost

An SBIC finances small firms in two general ways—by straight loans and by equity-type investments that give the SBIC actual or potential ownership of a minority of a small business' stock.   All financings must be for at least 5 years, except that a borrower may elect to have a prepayment clause included in the financing agreement.

Straight loans

While most SBIC's want an opportunity to share in the growth and potential profits of the small companies they finance, many SBIC's will make loans that involve no equity features. The small business that obtains a loan usually will be required to provide security, but this may take the form of a second mortgage, a personal guarantee or some other type of collateral that may not be acceptable to banks or other conventional lending institutions.

The interest rate on a loan is determined by negotiation between the SBIC and the small business, but is subject of course to the State's legal limit. Collateral requirements, terms of repayment and other parts of the loan agreement also are determined by negotiation.

How much may an SBIC invest

An SBIC may invest up to 20 percent of its capital in a single small business. For the smallest SBIC, the maximum loan or investment is $60,000; for the largest, it is several million dollars. In any event, several SBIC's may participate in financing the same small business and thereby increase the maximum investment.

An SBIC often will invest a negotiated amount in a small business and agree to advance additional funds after a specified period of time or after the small company has achieved pre-stated goals.

Management assistance

Since an SBIC's ultimate success is linked to the growth and profitability of its so-called portfolio companies—that is, those it has financed

—many SBIC's offer management services as a supplement to financing. These services sometimes are as valuable as the financing itself, although few small businessmen are aware of this when they first approach an SBIC.

Stocks and bonds

If you use stocks and bonds as collateral, they must be marketable. As a protection against market declines and possible expenses of liquidation, banks usually lend no more than 75 percent of the market value of high-grade stock.   On Federal Government or municipal bonds, they may be willing to lend 90 percent or more of their market value. The bank may ask the borrower for additional security or payment whenever the market value of the stocks or bonds drops below the bank's required margin.

What Are the Lender's Rules

Lending institutions are not just interested in loan repayments. They are also interested in borrowers with healthy profit-making businesses.   Therefore, whether or not collateral is required for a loan, they set loan limitations and restrictions to protect themselves against unnecessary risk and at the same time against poor management practices by their borrowers.   Often some owner-managers consider loan limitations a burden.   Others feel that such limitations also offer an opportunity for improving their management techniques.

Especially in making long-term loans, the borrower as well as the lender should be thinking of (1) the net earning power of the borrowing company, (2) the capability of its management, (3) the long range prospects of the company, and (4) the long range prospects of the industry of which the company is a part.   Such factors often mean that limitations increase as the duration of the loan increases.

What Kinds of Limitations

The kinds of limitations, that an owner-manager finds set upon his company depend to a great extent on his company.   If his company is a good risk, he should have only minimum limitations.   A poor risk, of course, is different.   Its limitations should be greater than those of a stronger company.

Look now for a few moments at the kinds of limitations and restric-

tions that the lender may set. Knowing what they are can help you
see how they affect your operations. The limitations that you will
usually run into when you borrow money are:

1. Repayment terms
2. Pledging or the use of security
3. Periodic reporting

A loan agreement, is a tailor-made document covering or referring
to all the terms and conditions of the loan. With it, the lender does
two things: (1) protects his position as a creditor—(he wants to keep
that position in as well a protected state as it was on the date the
loan was made)—and (2) assures himself of repayment according to
the terms.

The lender reasons that the borrower's business should *generate
enough funds* to repay the loan while taking care of other needs. He
considers that cash inflow should be great enough to do this without
hurting the working capital of the borrower.

Covenants—negative and positive

The actual restrictions in a loan agreement come under a section
known as covenants. Negative covenants are things that the borrower
may not do without prior approval from the lender. Some examples:
further additions to the borrower's total debt, non-pledged to other
of the borrower's assets, and issuance of dividends in excess of the
terms of the loan agreement.

On the other hand, positive covenants spell out things that the borrower must do. Some examples: (1) maintenance of a minimum
net working capital, (2) carrying of adequate insurance, (3) repaying
the loan according to the terms of the agreement, and (4) supplying
the lender with financial statements and reports.

Overall, however, loan agreements may be amended from time to
time and exceptions made. Certain provisions may be waived from
one year to the next with the consent of the lender.

You can negotiate

Next time you go to borrow money, thrash out the lending terms
before you sign. It is good practice no matter how badly you may
need the money. Ask to see the papers in advance of the loan closing.
Legitimate lenders are glad to cooperate.

Chances are that the lender may "give" some of the terms. Keep
in mind also that while you are mulling over the terms you may want

to get the advice of your associates and outside advisors. In short, try to get terms you know your company can live with. Remember, however, that once the terms have been agreed upon and the loan is made (or authorized as in the case of SBA), you are bound by them.

The Loan Application

Now you have read about the various aspects of the lending process and are ready to apply for a loan. Banks and other private lending institutions, as well as the Small Business Administration, require a loan application on which you list certain information about your business.

For purposes of explaining a loan application, we use the Small Business Administration's application for a small loan (SBA Form 6B)—one for $15,000 or less and maturities not exceeding 6 years. The SBA form is more detailed than most bank forms. The bank has the advantage of prior knowledge of the applicant and his activities. Since SBA does not have such knowledge, its form is more detailed. Moreover, the longer maturities of SBA loans ordinarily will necessitate more knowledge about the applicant.

Before you get to the point of filling out a loan application, you should have talked with an SBA representative, or perhaps your accountant or banker, to make sure that your business is eligible for an SBA loan. Because of public policy, it cannot make certain types of loans. Nor can it make loans under certain conditions. For example, if you can get a loan on reasonable terms from a bank, SBA cannot lend you money. The owner-manager is also not eligible for an SBA loan if he can get funds by selling assets which his company does not need in order to grow.

When the SBA representative gives you a loan application, you will notice that most of its 11 sections are self-explanatory. However, some applicants have trouble with certain sections because they do not know where to go to get the necessary information. Section 3—Collateral Offered—is an example. A company's books should show the net worth of assets such as business real estate and business machinery and equipment. "Net" means what you paid for such assets less depreciation.

If an owner-manager's records do not contain detailed information on business collateral, such as real estate and machinery and equipment, he sometimes can get it from his Federal income tax returns. Reviewing the depreciation that he has taken for tax purposes on such collateral can be helpful in arriving at the value of these assets.

If you are a good manager, you should have your books balanced monthly. However, some businesses prepare balance sheets less regularly. In filling out Section 6—"Balance Sheet as of_____, 19   , Fiscal Year Ends_____"—of the SBA loan application, remember that you must show the condition of your business within 60 days of the date on your loan application. It is best to get expert advice when working up such vital information. In some cases, your accountant or banker may be able to help you.

Again, if your records do not show the details necessary for working up profit and loss statements, your Federal income tax returns (Schedule C of Form 1040, if your business is a sole proprietorship or a parternship) may be useful in getting together facts for Section 7 of the SBA loan application. This section asks for "Condensed Comparative Statements of Sales, Profits or Loss, etc." You fill in the blocks appropriate to your form of business organization—corporation, partnership, or proprietorship—and attach detailed profit-and-loss statements.

Insurance

SBA also needs information about the kinds of insurance a company carries. The owner-manager gives these facts by listing various insurance policies. If you place all your insurance with one agent or broker, you can get this information from him.

Personal finances

SBA also wants to know something about the personal financial condition of the applicant. Among the types of information are: personal cash position; source of income including salary and personal investments; stocks, bonds, real estate, and other property owned in the applicant's own name; personal debts including installment credit payments, life insurance premiums, and so forth.

Evaluating the Application

Once you have supplied the necessary information, the next step in the borrowing process is the evaluation of your application. Whether the processing officer is in a bank or in SBA, he considers the same kinds of things when determining whether to grant or refuse the

loan. The SBA loan processor looks for:
1. The borrower's debt paying record to suppliers, banks, home mortgage holders, and other creditors.
2. The ratio of the borrower's debt to his net worth.
3. The past earnings of the company.
4. The value and condition of the collateral the borrower offers for security.

The SBA loan processor also looks at (1) the borrower's management ability, (2) the borrower's character, and (3) the future prospects of the borrower's business.

For Further Information

How much will you need

Have you worked out what income from sales or services you can reasonably expect in the first 6 months? The first year? The second year? Do you know what net profit you can expect on these volumes? Have you made a conservative forecast of expenses including a regular salary for yourself? Have you compared this income with what you could make working for someone else? Are you willing to risk uncertain or irregular income for the next year? Two years? Have you counted up how much actual money you have to invest in your business? Do you have other assets you could sell or on which you could borrow? Have you some other sources from which you could borrow money? Have you talked to a banker? Is he favorably impressed with your plan? Do you have a financial reserve for unexpected needs? Does your total capital, from all sources, cover your best estimates of the capital you will need?
REFERENCES: MA 105, *Watch your Cash; Term Loans in Small Business Financing in Marketers Aids Annual No. 2* (40¢ Supt. Docs.); *A Handbook of Small Business Finance* (30¢ Supt. Docs.)

Recommendation

Do not ask for money until you have world out the problems. Nothing is more impressive to potential investors than a product that gives definite evidence of probable sales appeal.

Do not overlook well-to-do people in your search for capital to begin manufacturing your invention. Particularly because of the time n sary to introduce a product until a profit is realized, inve returns would be considered capital gains and taxed at a lower rate than a wealthy person's normal income. A classified ad titled "Capital Wanted" may bring all the money you need to begin operations.

IX. Managing Your Business

Chapter VIII offers guidance in starting a business. But you are not ready to start your own business until you have made some study of the problems of managing it. What types of management problems will you face? Are you familiar with the buying techniques and markets for the materials and supplies you will need? Do you know how to price your merchandise? What are the best methods of selling them? Have you given thought to selecting and training the personnel you will need to help you run your business? Are you prepared to keep adequate records?

Inadequate and inefficient management causes more business failures than any other factor. There are few operators of small businesses who can afford not to make constant effort to improve management skills. Some of the problems will be touched upon here. But you should not limit your reading or investigating to these few pages. Through observation, conversation and reading you should continually search for better management methods that may be adapted to your particular business.

Buying and pricing

Skillful buying and pricing are important essentials of profitable operation. In the operation of your business you will be constantly called upon to answer the questions of what, from whom, when, and how much to buy.

Determining what to buy means finding out the type, kind, quality, brand, size, color, and style that will sell the best. This requires close attention to salesmen, trade, journals, catalogs, and any signs indicating the likes and dislikes of your customers. Analysis of your own sales

The authors wish to thank the U.S. Small Business Administration for permission to edit and publish portions of this chapter.

records is particularly helpful. Even the manufacturer should view the problem. Through the eyes of his customers before he decides what materials, parts, and supplies to purchase.

Determining from whom to buy involves locating suitable sources. You may buy directly from producers, from wholesalers who own the goods they sell, or through other middlemen who do not take title to the goods they sell. However, if your business is small, your choice probably will be restricted to the channel of distribution or type of market representative used by the producers whose goods you wish to purchase. You will select the suppliers that carry the kind and quality of goods you have determined as best suited to the needs of your customers.

As a policy, you may spread your purchases among many suppliers to gain the advantages of the most favorable prices. On the other hand, you may concentrate your purchases with as small a number of suppliers as possible. This policy gives you the advantages of

1. Receiving more attention and help from your suppliers, who know you are giving them most of your business.
2. Having a smaller inventory investment.
3. Having larger purchase orders which may permit larger discounts.
4. Simplifying your credit problems.
5. Maintaining a fixed standard in your products, if you are buying materials to be used in making other goods.

In spite of arguments in favor of placing orders among many suppliers, it is usually better for the small business to concentrate its purchases and work closely with a few.

The question of when to buy deserves attention in the many businesses having seasonal variations in sales volume. More merchandise for sale must be in stock prior to the seasonal upturn in sales volume. As sales decline, less merchandise is needed. This means that purchases of goods for resale and materials for processing should vary accordingly.

Also, speculative buying is closely related to the question of when to buy. Should a buyer stock up when he believes prices are unusually low in order to take advantage of a possible price rise? Ordinarily, you should avoid speculative buying because it interferes with the normal operations of your business. While inordinate profits may sometimes be made, substantial losses are equally likely.

How much to buy should be answered by your own records as soon as you have had enough experience to judge. Do not overbuy. This will lead to serious financial trouble. On the other hand, you cannot sell merchandise if you do not have it.   Careful analysis of your records plus good judgment will determine the quantities you should buy.

To help with these problems of buying, you should keep some records.   Set up some system of stock control.   Stock control is a method of keeping stock in balance—neither too large nor too small— with a proper proportion and an adequate assortment of sizes, colors, styles, qualities, and so on.   It provides a guide to tell you what, from whom, when, and how much to buy.

Much of your success in business will depend on how well you price your goods or services.   If your prices are so low that your margin does not cover expenses, or so high that you can not build up sales volume, you will fail to make profits.

Before you open your business you will have decided upon the general level you expect to maintain—that is, whether you expect to cater to people buying in the high, medium, or low price range.

After establishing the general price policy, you are ready to price individual items.   To be certain that you do not underprice, you should know the percentage of gross margin to sales needed in the total of all items to cover expenses and profit.   If you have been in business a year or longer, you can analyze your past records and find out the percentages for operating expenses and net profit.   If you are just starting a business, you will have to estimate your sales and expenses carefully.   Suppose you figure the margin you need to be 30 percent of sales.   To obtain a 30 percent gross margin on an item you must mark up its cost to you by 42.9 percent.   This is because margin is a percentage of sales, while markup is a percentage of cost of merchandise.   Many wholesalers furnish tables to retailers showing markup percentages on cost price for different margin percentages of sales price.

However, in this example, you would not obtain 30 percent margin if each individual item were marked up only 42.9 percent.   No allowance has been made for markdowns and shrinkage in this markup. Markdowns are reductions from the original selling prices.   Among the reasons for them are overstocking as a result of unwise buying, sudden changes in style, unseasonable weather, soiled and faded goods used for displays, and broken lots, odds and ends, and odd pieces left at the end of the season.   "Shrinkage" is the term used for losses due to theft, spoilage, breakage, etc.   Allowances for markdowns and shrinkage must be added if you expect to maintain an average markup of 42.9 percent, or, to put it another way, a margin of 30 percent.

In manufacturing a product, the costs of direct labor, materials and supplies for production, parts from other concerns, plant overhead, selling and administrative expenses must be carefully estimated. Such a method of pricing implies a knowledge of the unit cost of each product you make. If your business is small you may find it neither practicable nor necessary to keep detailed records of unit costs. But as the business grows it becomes more difficult for you, as the owner, to know sufficiently well the costs of materials and labor going into each product or service. A method of determining unit costs with some degree of accuracy usually becomes desirable.—It may be wise, when that time comes, to consult a public accountant about this. If you plan to operate a small factory, a good reference is: *Cost Accounting for Small Manufacturers*, Small Business Management Series.

Pricing

Consult Chapter IV on Marketing.

Selling

When you operate a factory, you will have to sell. Probably, there is no other business function about which so much has been written. For that reason little is presented here.

Your direct methods of selling are through personal sales efforts, advertising and, for most businesses, display (including the packaging and styling of the product itself). The last named subjects are discussed in *Design Is Your Business*, Small Business Management Series. Establishing favorable relations through courtesy and special services and a good reputation with the general public are indirect methods of selling. While the latter should never be neglected, in this brief discussion comments will be confined to the direct methods of selling.

To establish your business on a firm footing requires a great deal of aggressive, personal selling, except possibly in a mail-order business. You will have well-known competition to overcome. Or, if your idea is new enough to minimize competition, you have the extra problem of convincing people of the value of the new idea. Personal selling work is almost always necessary to accomplish this. If you are not a good salesman, seek an employee or associate who is.

Another method of building sales is by advertising. This may be done through newspapers, trade papers, the classified section of the telephone directory, and other published periodicals; radio and television; handbills, and direct mail. The media you select, as well as the message and style of presentation, will depend upon the particular group you wish to reach. Do not advertise unless you plan and prepare the advertising carefully. Otherwise, it will be ineffective. This method of selling can become highly expensive and it is wise to place a limit upon the amount you plan to spend, then stay within that limit. To help you in determining how much to spend, study the operating ratios of similar businesses. You may find it wise to employ the services of an advertising agency. *"How Advertising Agencies Help Small Business,"* Chapter 18 or *Management Aids for Small Business: Annual No. 2* contains information about what agencies can do and the costs of their services.

A third method of stimulating sales is effective display. If you have had no previous experience in the art of display, you will want to study the subject. You should observe displays of other businesses and read books, trade magazines, and the literature supplied by equipment manufacturers. It may be wise to hire a display expert or obtain the services of one on a part-time basis.

The proper amount and types of selling effort to use vary from business to business. What is effective in one business may be bad taste in another. In any event, this business function will consume a considerable amount of your time and effort.

While this important function deserves a great deal of attention, do not over-emphasize it at the expense of long-range objectives. If so much time is devoted to selling effort that you neglect records and lose sight of your plans for profit, disaster may result. Many retail operators, for example, have stimulated high sales volume while failing to record markdowns as they were taken. At the end of the season, shortages due to unrecorded mark-downs have been known to wipe out net profit.

Selecting and training personnel

If your business is going to be large enough to require outside help, one of the first jobs will be to select and train one or more employees. You may start out on a small scale with only members of your family or business partners to help you. But if the business grows (as you hope it will) the time will come when you must select and train personnel.

Careful choice of personnel is essential to protect the reputation of your business. To select the right employee, you should plan beforehand what you want him to do, and then look for the applicant to fill your particular needs.   Often one of the major mistakes in choosing an employee is to hire him without a clear knowledge beforehand of exactly what you want him to do.   It is true that in a small business you will need flexible employees who can shift from task to task and who may be called upon to perform unexpected tasks.   Nevertheless, you should plan your hiring to assure an organization capable of performing every essential function.   Write down the job descriptions. For example, you should answer such questions as these before hiring: Will a sales person also do stockkeeping or bookkeeping?   Will the laborer have sales duties in the absence of the manager?

After you have written down the duties, look for an applicant who can perform them.   It is better to seek and select rather than wait for applications.   Some sources of new employees are:

1.   Suggestions of friends, business acquaintances, employment men, and others.
2.   Your nearest United States Employment Service office.
3.   Placement bureaus of high schools, business schools, and colleges.
4.   Employment agencies, the YMCA, the YWCA, and similar sources.
5.   Want ads in local newspapers.
6.   The voluntary applicant.

An employee well selected is only a potential asset to your business. Whether or not he becomes a real asset depends upon the way you train him.   Much faulty training may be avoided if you remember

1.   To allow sufficient time for training.
2.   Not to expect too much from the trainee in too short a time.
3.   To have the emplyee learn by performing under actual working conditions, with close supervision.
4.   To follow up on your training.

An illustration of the importance of these points in a retail store is the story of a salesgirl who was trained in a special training session. Methods of filling out sales checks and operating the cash register were not only explained and demonstrated, but were taught by having her perform hypothetical sales transactions.   Location of stock and points about the merchandise were explained.   Then the girl was placed on the selling floor.   During the noon hour she was left alone in one section.   Customers swarmed around her asking questions and demanding service.   She became more and more confused, forgetting what she had learned under quiet, artificial conditions.   In desperation, she walked off the floor and never returned.

While the merchant in this case thought he was giving attention to training, he was actually violating the rules enumerated here. First, he had not allowed sufficient time for training. Second, he expected too much to be absorbed by the trainee in too short a time. And finally, he had neglected to continue the training on the job under close supervision.

You must check the employee's performance after he has been on his own and has had practice. Re-explain key points, suggest knacks or short cuts, bring him up to date on new developments, and encourage him to ask questions. Training is a continuous process.

You may think that too much care in selecting and training an employee is unnecessary. After all, if he doesn't prove satisfactory you can fire him, you will say. But remember that selecting an employee, training him, and breaking him in as a member of your organization is an expensive process. Besides the cost of the time involved, this expense is hidden in a number of other costs, such as wasted supplies, damaged equipment, and low quality work. And, most important, if your employee is unsatisfactory, your greatest expense may be that of overcoming customer ill will and a poor business reputation.

Other management problems

You will be faced with other types of problems besides those concerned with buying, pricing, selling, and personnel. For example, if you extend credit to customers, you will become interested in the proper management of credits and collections; as a retailer or wholesaler, you may have delivery problems; as a manufacturer, efficient production is a major factor. Each of these demands special treatment which cannot be described in a single chapter.

Importance of adequate records

Studies of business failures show that reasons for failure can frequently be attributed to inadequate records. Absence of records is not itself the cause of difficulties, but it accounts for the businessman's inability to see in advance the direction in which he is going. With up-to-date records, he may foresee impending disaster in time to take steps to avoid it. While extra work may be required to keep an adequate set of records, this work will more than repay you for the effort and expense. If you are not prepared to accept this chore, you should not try to operate your own business. You need records to substantiate:

1. Your returns under Federal and State tax laws, including income tax and social security laws.
2. Your requests for credit from equipments manufacturers or a loan from a bank.
3. Your claims about the business, should you wish to sell it.

But most important to you, you need them to help increase your profits. With an adequate, yet simple, bookkeeping system you can answer such questions as these:

1. How much business am I doing?
2. What are my expenses?   Which expenses appear to be too high?
3. What is my gross profit margin, my net profit?
4. How much am I collecting on my charge business?
5. What is the condition of my working capital?
6. How much cash do I have on hand and in the bank?
7. How much do I owe my suppliers?
8. What is my net worth—that is, what is the value of my ownership of the business?
9. What are the trends in my receipts, expenses, profits, and net worth?
10. Is my financial position improving or growing worse?
11. How do my assets compare with what I owe?   What is the percentage of return on my investment?
12. How many cents out of each dollar of sales are net profit?

These and other questions may be answered by preparing and analyzing balance sheets and profit-and-loss statements. For guidance about how your financial statements can be interpreted for management purposes, read *Ratio Analysis for Small Business*, Small Business Management Series.

Q. What are the other alternatives besides producing the product yourself?

A. An inventor can have the product made by a manufacturer completely or the parts manufactured with the assembling, finishing and shipping handled by the inventor.

Q. What are the other alternatives besides marketing the product yourself?

A. An inventor can turn over part or entire production to a nationwide distributor, or appoint regional franchised dealers.

Are you qualified to supervise buying and selling   Yes   No

Have you estimated your total stock requirements   —   —

Do you know in what quantities users buy your product
or service? — —

Do you know how often users buy your product or
service? — —

Have you made a sales analysis to determine major
lines to be carried? — —

Have you decided what characteristics you will require
in your goods? — —

Have you set up a model stock assortment to follow
in your buying? — —

Have you investigated whether it will be cheaper
to buy large quantities infrequently or in small
quantities frequently? — —

Have you weighed price differentials for large orders
against capital and space tied up? — —

Have you decided what merchandise to buy direct from
manufacturers? — —

Will you make your account more valuable to your
suppliers by concentrating your buying with a few
of them? — —

Have you worked out control plans to insure stocking
the right quantities? — —

REFERENCES: MA 120, *Checking Your Marketing Channels;* MA
123, *Getting the Most From Your Purchasing Dollar;*
SM 28, *Profitable Buying for Small Retailers;* SM
56, *Advertising for Profit and Prestige;* SM 60, *Sales
Promotion Pointers for Small Retailers;* SBB 37,
Buying for Retail Stores.

Looking Into Special Requirements

You are not ready to start your business until you have considered
special requirements in connection with your proposed new enterprise. You must become generally familiar with the kinds of legal,
tax and insurance problems you will face; for example: What laws
and regulations will affect you? To what taxes will your business be
subject? How many kinds and how much insurance should you
carry? Are there other special requirements pertaining to the particular line of business you propose to enter?

Laws and regulations

The more common types of laws and regulations are reviewed briefly here under the headings licensing, regulation of trade practices, and labor relations. The information given here is not intended as a substitute for legal advice nor can it be considered as making it unnecessary for you to obtain such advice. The services of a competent person should be sought if you require legal assistance.

Licensing

Licensing controls directly affect practically all businesses. The degree of regulation, however, will vary, depending upon the type and location of the enterprise. If the operations are intrastate you will be concerned primarily with State and local, rather than federal licensing. If you are engaged in interstate commerce, federal regulations become increasingly significant.

Most licenses require payments of fees and are usually issued on an annual basis. Ordinarily, as a prerequisite to the issuance of a license, it is necessary to file a written application. State, municipal and country authorities should be contacted for complete information regarding license requirements. These authorities should also be consulted with respect to regulations governing the construction and use of business premises, including building codes, zoning restrictions, and health and safety regulations.

Regulation of trade practices

Some business practices are prohibited or restricted by federal and state legislation designed to encourage competition. The federal laws govern dealings in interstate commerce, while the state legislation regulates transactions in intrastate. Occasionally, some of the laws are amended, and new interpretations are made by the courts. Your lawyer and your Chamber of Commerce or business association should be good sources of guidance on how such laws or proposed laws may affect you.

Labor relations

Federal and state legislation affecting employer-employee relations deals with the settlement of labor disputes and with wages, hours, and working conditions.

The Labor Management Relations Act, as amended by the Labor-

Management Reporting and Disclosure Act of 1959, applies to all employers and employees engaged in industries affecting commerce between the states. This act guarantees the right of workers to organize and bargain collectively with their employers, or to refrain from such activities. So that these rights may be exercised, it places certain limits on the activities of employers and labor organizations. In addition to this act pertaining to interstate commerce, some states have enacted laws designed to encourage collective bargaining and to define unfair labor practices.

Federal wage and hour legislation is contained in the Fair Labor Standards Act, as amended. This act establishes minimum wage rates for employees engaged in interstate or foreign commerce and for other employees in certain kinds of enterprises whose activities are related to interstate or foreign commerce. The act further provides that most such employees shall be paid time and one-half for work in excess of a specified number of hours per week. It covers employees working in the 50 states and in the District of Columbia, Puerto Rico, the Canal Zone, Guam, Samoa, and the Virgin Islands. Employment of minors under 16 years of age, with certain exceptions for nonmanufacturing occupations, is also prohibited by the act.

In addition to the Fair Labor Standards Act, the Walsh-Healey Public Contracts Act, the Davis-Bacon Act, and other related acts establish wages, hours, and working conditions applicable to government contractors.

Many state statutes affect employer-employee relationship. These Statutes deal with such problems as the health and safety conditions under which employees work, workmen's compensation and unemployment insurance, child labor regulation, and special laws affecting women.

To conclude, it is well to check with local, state and, if you expect to engage in interstate activities, federal authorities regarding laws covering your operations, whatever your business.

Taxes

Before establishing your enterprise, some time should be devoted to the exploration of the tax situation. Your business will be subject to federal, state, and local taxes. Among the federal taxes for which you may be liable are social security taxes (shared by you as employer and your employees), excise taxes, and, if your business is incorporated, the corporate income tax. From your employees' wages you must deduct their share of the old-age and survivor's insurance, and, in cases, federal unemployment compensation contributions, as well as withhold an account for their current payments of their individual income tax.

If you are an employee of your own corporation, the withholding provisions of the social security and individual income taxes also apply to you.   The 1954 amendments to the Social Security Act extending coverage of the act to those self-employed.   Now, the owner of a small business concern can build for himself retirement rights and benefits under the act.

If you are a sole proprietor or partner, your personal income tax payments must be prepaid or kept current on a quaterly basis.   Go to the local office of the Director of Internal Revenue for information about your federal tax obligations.   An excellent booklet on this subject is *Tax Guide for Small Business*, Internal Revenue Service.   This may be purchased from the Superintendent of Documents, U. S. Government Printing Office, Washington 25, D. C.

Because state and local taxes differ greatly from place to place, it is only possible to suggest the general nature of these taxes and local sources of information.   All 50 states, as well as the District of Columbia and Puerto Rico, have state unemployment compensation taxes, although they are not entirely uniform.   Information concerning this tax should be obtained from the unemployment compensation agency in the state where you will be doing business.

The other more common types of taxes levied by the states are income, property, sales, and occupation or business license taxes.   Information concerning the state taxes and fees which apply to your particular business can be obtained from the State agency responsible for tax and revenue collections in your state.

Many municipalities have property taxes and one or more forms of taxes upon business, usually license taxes.   Also sales and income taxes are levied by municipalities.

No specific advice can be given in connection with local taxes other than to consult your county and local tax officials—the treasurers or tax collectors—for tax information as it relates to your business in your particular locality.

Insurance

Before opening your business you should arrange for adequate insurance protection, otherwise a part or all of your investment may be lost.   Although you plan to take extensive precautions against damage or loss occurring to your physical property, you will still need insurance. Protection against this kind of loss is afforded by such types as fire, lightning, and windstorm insurance, use and occupancy insurance, robbery and burglary insurance, and casualty insurance. To protect against claims for personal injury are public liability insurance and workmen's compensation or employer's liability insurance. In most states an employer is required to carry the latter. Life insurance on partners or corporate officers whose personal services are regarded as essential to the welfare of the business has become common practice in recent years. Other types of insurance such as fidelity and surety bonds may be needed. More information on business insurance is published in Chapters 10, 11, 12, and 13 of *Management Aids for Small Business: Annual No. 1.*

The subject of proper insurance coverage is an involved one and it is wise to consult a reliable insurance agent, broker, or company representative for advice.

What laws will affect you?   Yes   No

Have you investigated what, if any, licenses to do business are necessary? — —

Have you checked the health regulations? — —

Are your operations subject to interstate commerce regulations? — —

Have you seen your lawyer for advice on how to meet your legal responsibilities? — —

    REFERENCES: MA 108, *Selecting a Lawyer for Your Business;* SM 42, *FTC and Guides Against Deceptive Pricing; Small Business and the Federal Trade Commission: Marketers Aids Annual 2* (40¢ Supt. Docs.)

What other problems will you face?

Have you worked out a system for handling your tax requirements? — —

Have you arranged for adequate insurance coverage? — —

Have you worked out a way of building a management team? — —

Does your family (if any) agree that your proposed venture is sound? — —

Do you have enough capital to carry accounts receivable? — —

Will you sell for credit? — —

Have you worked out a definite returned goods policy?　—　—
Have you considered other management policies which
　must be established?　—　—
Have you planned how you will organize and assign the
　work?　—　—
Have you made a work plan for yourself?　—　—
　　REFERENCES:　MA 103, *Organizing the Owner-Manager's Job;* MA
　　　　　　113, *"Tailor-Make" Your Excutive Staff; Building
　　　　　　Sound Credit Policies for Small Stores; Marketers
　　　　　　Aids Annual I* (45¢ Supt. Docks.); SM 49, *Improving
　　　　　　Collections from Credit Sales; Business Insurance;
　　　　　　Management Aids Annual No. I* (65¢ Supt. Docs.);
　　　　　　*How Good Records Aid Income Tax Reporting;
　　　　　　Management Aids Annual No. 3* (45¢ Supt. Docs.);
　　　　　　*Appeal Procedure for Income Tax Cases; Manage-
　　　　　　ment Aids Annual No. 4* (45¢ Supt. Docs.)

Will you keep up to date?

Have you a plan for keeping up with new developments
　in your line of business?　—　—
Have you a small group of qualified advisors from whom
　you can get help in solving new problems?　—　—
　　REFERENCES:　MA 117, *Selecting Marketing Research Services;*
　　　　　　MA 125, *Building Growth-Mindedness Into Your
　　　　　　Business;* SM 54, *Store Modernization Check List.*

Steps in Making a Business Decision

Making effective decisions is the most important part of your job
as a small marketer.　You have to decide what is to be done, when,
why, and by whom.　Some small marketers use a logical system for making decisions.　Other small business owners try to be systematic but
often overlook pertinent facts.　And, of course, there are a few small
business owners who "fly by the seat of their pants."　These men often
trust to luck in making their decision.

Seven steps for sound decisions

Seven logical steps can help you to rule the element of luck out of
your business decisions.　These seven steps are:
1.　State the problem.

2. Determine goals and objectives.
3. Determine and analyze the factors bearing on your goals and objectives.
4. List possible solutions or courses of action.
5. Compare the possible solutions.
6. Choose the most logical solution.
7. Determine the specifics to fulfill the choice you made.

As you work through these seven steps, talk with your outside advisers, such as your accountant, commercial banker, attorney, real estate broker, and suppliers. These people and other advisers can provide specialized advice and help which should be useful to you as it was to John Boldso when he used this seven-step process to make a major business decision.

1. State the Problem. John Boldso's problem concerned growth and the possibility of relocating his firm, Boldso Sales and Repair Company. With the help of 15 employees, he sold and repaired household appliances. The decision he felt he needed to make was: Should I stay in this location? (He had started his firm there 20 years ago.) Or relocate? Should I continue to sell appliances? Or should I do only repair work?

Changing conditions made Mr. Boldso realize that his firm could not drift with events. His potential market for retail sales was limited because his location was in a declining area in the center of the city. Added to this trend was the fact that an expressway was to be built near his firm. It would cause further drastic changes in the traffic that passed Mr. Boldso's store.

Mr. Boldso spelled out his problem in detail by looking at the three parts involved. He took an objective look (1) at himself, (2) at his firm, and (3) at the pertinent outside factors.

Mr. Boldso. He started by evaluating himself as an individual. He was 50 years old, his health was good, and he knew that he wanted to keep on working for himself. He was convinced that his experience and management knowledge would enable him to handle a larger business if he decided to expand.

Mr. Boldso then looked at his own financial requirements. They were around $12,000 per year, but he knew that he would need more during the next 8 years in order to send his children to college. M of his financial resources were tied up in his business so that th earnings would have to increase in order to provide Mr more income.

His Firm As An Enterprise. After looking at himself, Mr. Boldso examined his firm. It had a reputation for efficient service so that many of his customers would continue to deal with him even if he moved. His employees were loyal and well-trained. However, none of them were managers.

All of the management decisions were made by Mr. Boldso, himself. He supervised all phases of the business. He had not provided for management succession, and his 18-year-old son did not seem interested in the business. These factors meant that if Mr. Boldso wanted to expand he would have to hire some managerial personnel.

The firm also would need additional working space, storage space, and equipment repair facilities whether it moved or remained in its present location. The firm had the financial resources for making such changes at the present location. Moreover, Mr. Boldso could get commercial bank credit for expanding or moving to another location. His present building and land was worth $30,000. An adequate building in another location would mean an investment of about $120,000.

With these facts in mind, Mr. Boldso looked at growth prospects for his firm. He reasoned that growth would be hard to achieve in his present location. Although his growth rate had been good over the years, it had slowed down recently. His excellent relations with customers, suppliers, and other business associates would not make up for the run-down neighborhood. Some women customers from outlying residential areas would stop trading in this "tough" neighborhood. Competition was no problem in his present location, and Mr. Boldso could probably retain his competitive advantage—doing excellent repair work—if he moved to another spot in the metropolitan area.

External Factors. Mr. Boldso looked at the external factors that were pertinent to his problem. He knew the date when expressway construction would start near his store. This construction did not favor his present location because it would reduce automobile and pedestrian traffic on his street.

Another factor that Mr. Boldso had to consider was rising land costs. Other businesses would be displaced by the expressway, and he already knew that the cost of land was increasing in the desirable outlying business districts.

2. *Determine Goals and Objectives.* Mr. Boldso then made a list of the goals that he wanted for himself and for his business. His long-range goal was business growth so that his own income would increase to about $16,000 a year. He wanted to achieve this goal in the same

line of business and in the same general area.   He and his family preferred to live in their present community.

He listed the following needs that his business would have to provide
for in the near future.   He called them short-range objectives.

1.   Improvements in his facilities were needed even if he remained
in his present location.   (In fact, they were overdue.)

2.   The firm needed a manager to head up its largest activity—the
repair department.

3.   The firm needed someone who could relieve Mr. Boldso of some
of the administrative details.   (This assistant, and perhaps the repairs
manager, would have to be hired from the outside.)

4.   The firm needed to reduce its retail sales activities in order to
take care of its excellent accounts with large department stores for
warranty repairs.

5.   The firm should be reorganized so that its actual management
did not depend entirely on Mr. Boldso.

3.   Determine and Analyze Factors Bearing on Goals and Objectives.
In the third step, Mr. Boldso appraised many of the same factors he
had looked at in Step I (when he stated his problem).   He needed
to know, as nearly as possible, the effect these factors could have on
his personal and business objectives.

Among the favorable factors were those relating to Mr. Boldso as
an individual.   Age, health, experience, business knowledge, financial
resources, and relations with others were all favorable.   His willingness to consult with outside advisors, such as commercial bankers, real
estate brokers and accountants, could also work to his advantage.
Most of the factors relating to his firm as a business enterprise were
also favorable.   However, there were four areas that presented problems.   They were:

1.   The shortage of management personnel within the firm.

2.   Increased costs.   Upgrading his present facilities or moving to
another location would increase the firm's building occupancy costs.
However, on the favorable side was the fact that suitable space was
available for relocation.

3.   Any change in the firm's present way of doing business would
increase operating expenses and invested capital.

4.   If changes were made, Mr. Boldso would have to expect a lower
return on invested capital until the firm's volume increased.

He reasoned that over the long pull, the economic climate of the
metropolitan area would be improved by the expressway construction.

However, he knew that he would lose business to outlying shopping areas as people moved to the suburbs. So, he reasoned that prospects were good for a well-managed household appliance repair service.

4. *List Possible Courses of Action.* . In the third step Mr. Boldso, in effect, made a study of the possibilities that were open to him. As he saw it, there were four possible courses of action:

1. He could remain in his present location and sell and repair appliances.

2. He could move to a more suitable location in an outlying area.

3. He could stay where he was, concentrate on repairs, and set up pick-up stations near several outlying shopping centers.

4. He could liquidate and reinvest in something quite different.

5. *Compare the Possible Solutions.* In order to determine the best course of action, Mr. Boldso compared the four possibilities. First, he ruled out No. 4, and then he weighed the advantages and disadvantages of each of the three remaining.

Remaining in his present location with improvements had the advvantages of less invested capital and lower operating costs. It was also a centralized location. On the other hand, growth—increased sales volume and increased profits—would be hard to achieve at his present location.

Moving to an outlying area and enlarging his facilities offered the possibility of growth. Yet such a move could cause some inconvenience to some of his present customers, such as hospitals, and hotels. He could expect to get more new customers from the suburbs, but his building occupancy costs would be significantly larger.

One of Mr. Boldso's competitive advantages was his complete inventory of replacement parts for all major lines of household appliances. This inventory required a large investment. It was a strong reason for doing business in only one location. Another good reason for only one location was that he had only two well-trained sales-clerks. These men could quickly determine for the customer (1) whether parts were available, (2) what the cost of repairs would be, and (3) when the repairs would be completed. This excellent customer relations advantage would be hard and expensive to duplicate in more than one location in a short period of time.

Concentrating on repair work and setting up pickup stations in outlying areas involved the greatest risk of invested capital. However, such action also offered the largest opportunity for growth. One problem would be the shortage of management personnel. In two or more locations, Mr. Boldso would have to hire several managers and make some personnel shifts.

6. *Choose the Most Logical Solution.* With the advantages and disadvantages of the three possibilities before him, Mr. Boldso was ready to choose the most logical solution to his problem.

First, he decided to stay in his present location. Next, he decided to stop selling appliances and concentrate on repair work. He would ask for more warranty repair business from the large appliance stores. He would enlarge his building and improve his facilities so that he could handle the work more efficiently. He also decided to experiment by setting up pick-up and delivery facilities in three major shopping centers.

7. *Details for Carrying Out Decisions.* Mr. Boldso's next step was that of determining the details that were needed in order to carry out his decision. First, he would need financial assistance. He already knew that his bank would help him.

Next, Mr. Boldso discussed his building enlargement plans with an architect. He also contacted a real estate broker to help him find rental space for his pick-up and delivery stations. Mr. Boldso arranged to buy a panel truck for picking up work from the branches and for delivery service to his big customers in the city.

A very important detail was that of better management tools. Mr. Boldso asked his accountant to develop new systems and procedures that would help him in managing the enlarged business.

In personnel, Mr. Boldso needed three men as managers of his pick-up stations. He trained them at his main location. Over the long pull, Mr. Boldso needed (1) someone to help him with the new administrative details and (2) someone to manage his repair service department. However, he decided to postpone action on these two persons. On the long-range goal of re-organizing his firm, Mr. Boldso took his banker's advice—change the firm from a sole proprietorship to a closely held corporation. Mr. Boldso began to discuss the possible advantages of such a change with his attorney.

Using outside advisers

As you use the seven steps for making a business decision, keep in mind the importance of talking with outside advisers as did Mr. Boldso. As he studied the various parts of his problem, he talked, for instance, with his commercial banker about financial assistance.

Among other outside people with whom Mr. Boldso talked were an architect, real estate broker, an accountant, and city planners. All of them gave him assistance in specialized areas that were beyond his immediate knowledge.

Problems are the same

The seven-step procedure for making a business decision can be used by the small business owner regardless of the size of his firm. The underlying problems of the two-person shop are the same as those of a bigger store. For example, both types of firms have to find answers to the same questions when making decisions about merchandise: What kinds of new items can I carry? Where can I get these items? How much stock should I carry? How will I pay for it?

Try using the seven steps. Like any management tool, they can help to make your job easier.

Use of Outside Information in Small Firms

Most formal business information is distributed by six major sources: (1) trade associations, (2) the trade press, (3) professional consultants, (4) public libraries. (5) business and commercial schools, and (6) governmental agencies. Suppliers also furnish a great deal of information about their products.

Trade associations

Trade and business associations throughout the nation are one of the most active and complete sources of business information. Many businessmen look on these associations as their most important sources of information. Analysis of materials published by trade associations showed that approximately one-third of the associations published how-to information. A little more than half offered basic data.

The trade press

The term "trade journals" is used here to mean periodicals planned for businessmen in a particular function or line of business. (Some are published by trade associations, but the majority originate with independent publishing firms.)

The Predominance of Periodicals. Nearly all businessmen (90 percent) read one or more trade journals. Thus, it is apparent that periodicals represent the bulk of the business literature received by the small businessmen interviewed. This predominance of periodicals is streng-

thened by the fact that they are received again and again through the
year, while the nonperiodicals are received only once.

Most businessmen read trade journals to keep informed about new
developments and about what is going on in their lines of business.
Other business publications are read to keep up with trends in general
business conditions.

Professional consultants

Many small businessmen cannot afford to have full-time specialists
and professional people on their staffs, but are able to absorb the costs
of professional advice on a short-term or one-time basis. Specialists
available for this sort of help include management consultants, industrial engineers, business-school professors and instructors, accountants,
and so on.

Public libraries

Few public libraries originate information, but all are important
points of distribution for information that would not otherwise be
conveniently available to small businessmen. Libraries throughout
the country are giving increased attention to developing and publiciz-
ing their resources for business and industry.

Business and commercial schools

Information from business and commercial schools is available
through five channels: (1) extension classes, (2) special seminars and
short courses, (3) management development programs, (4) correspondence courses, and (5) published materials.

NASA

Technology is accumulating so rapidly that small businessmen need
to be alert to many sources of research and development information.
One such source is the space-related technology being generated by
the National Aeronautics and Space Administration (NASA). Some
of the data can be adapted to industrial use in small manufacturing plants.

Part of NASA's work is that of keeping small business owners, and
others, abreast of such R&D opportunities.   This task is done by
NASA's Office of Technology Utilization (OTU). It consists of seeking
out space technology with potential industrial uses, such as operating
techniques, management systems, materials, processes, products, and
design procedures.

Selling by Mail Order

Mail-order selling is an operation, rather than a kind of business, and is used by many organizations in trade and industry. Some firms obtain most or all of their business from mail orders, while others use this operation incidentally.

Selling by mail is often thought of as an easy way to make money in one's spare time. Stories and advertisements give the impression that several hundred dollars or more can be made by using little effort to sell various kinds of merchandise. The truth of the matter is that selling by mail is hazardous for the inexperienced person. So much organization is required, even for a part-time business, that the beginner finds himself spending many hours of his spare time with little or no return.

The best way to approach the mail-order business then is as a hobby or avocation with no immediate prospect of profit. To avoid office costs, the best procedure is to operate from the home. Ignorance of business problems, or expectation of high profits in the early stages will lead to discouragement. If a person acquaints himself with the rules, has a little capital, and is content to build a business slowly, he can be successful.

Several important pieces of information are needed by the beginner. The major principles are set forth in logical order are:

The market

One of the most important principles of mail-order selling is to know the market—that is, what kinds of persons will buy what kinds of merchandise in this way. Some goods do not lend themselves to distribution by this method, while others are naturally adapted to it. Some types of goods that can be sold by mail-order are: Novelties, Hobby items, Convenience goods, Lower-priced goods.

The mailing list

Building a list and maintaining it are two of the most important operations in mail-order selling. If a list is failing to reach the right kind of prospect, even an adequate line of merchandise will not sell. A list must be accurate and be revised constantly to perform its function efficiently. Several methods can be used for building lists:

1. *Keying advertisements.* A most successful method is to offer some inexpensive item that can be secured by filling in a coupon printed

within an advertisement. A number or letter may be placed inconspi-cuously in the coupon, or a fictitious department number or letter may appear in the address. If the advertisement is run several times, the number of responses from each insertion can be counted, and the effectiveness measured. From the cost of the advertisement itself in relation to the number of inquiries received, the seller can determine the cost per inquiry and the cost per sale. He can thereby judge whether or not his promotional costs are in line with his estimates and with the prices he is charging.

2. *Purchasing or renting lists.* List houses can furnish names and addresses of prospects for a product. They are purchased at so much per name or per thousand; the more selective the list is, the higher the cost per name. It is therefore important to assess the market in terms of the selectivity desired; otherwise, a too selective list may fall short of the anticipated volume of sales. Lists are always accurate when supplied, though there is no guarantee that they will remain so. The mobility rate of the population is very high, especially in urban areas (estimated at 15% to 20% annually); and a continual depreciation in accuracy of lists will cut into returns.

The return from an unselected list usually runs about 2 percent, whereas a selected one might yield as high as 5 to 10 percent. Although one can depend upon a 2 percent return from an accurate, unselected list, he can also expect as high as 10 percent error in the list within a year. If the basis for computing the return is reduced from 100 percent to 90 percent accuracy, the same return will then be 2 percent of the 90 percent, or 1.8 percent. This 0.2 percent can mean the difference between a profitable and an unprofitable venture.

If the list is performing as expected, its accuracy must be maintained by eliminating "dead" prospects and adding new ones from supple-mentary lists or advertising.

Maintaining the lists

As the cost of postage is increasing, the dealer must be alert to changes in address as they occur. It is important that correct initials, spelling of last name, and street and number are used. The zip codes by states and cities are now in effect. Zone numbers have been absorb-ed into the zip code numbers. Their use will help to speed delivery. Several classes of mail are available, some of which provide for return of undelivered matter:

1. *First Class.* All mail of this kind is sealed. It will be forwarded

if a new address has been furnished to the Post Office.   If a return
address is provided, any undelivered mail is returned to the sender.
Since first class is expensive, only more elaborate or personalized
pieces should be sent this way; sometimes it is used for mailings to a
highly selective list of prospects.

2. *Third or Fourth Class (Parcel Post).* As of January 10, 1962, procedures for undeliverable mail have been changed. Form 3547 has
been discontinued and replaced with a new uniform endorsement called
Return Requested.   These words are printed on the envelope or package
of any mail the sender wishes returned if undeliverable. The mail
will be returned showing new address or other reasons for nondelivery.
All undeliverable third and fourth class mail returned to the sender
will be charged at the applicable single piece rate or 8 cents per piece,
whichever is higher.

3. *Method of correcting lists.* Post offices will correct mailing lists
if individual names and old addresses are submitted on cards, at the rate
of 5 cents per card.   A minimum of $1 is charged.   Cards should be
about the size of a post card, one name to each, with sender's name
in the upper left-hand corner.

Rules and regulations

Before beginning a mail-order business, one should become acquainted with the various federal, state, and local laws and regulations
applicable.

Expanding Sales Through Franchising

Franchising offers small companies a way to accelerate their growth
—to increase sales.   In this method of distribution, the small manufacturer can obtain new outlets for his products by granting franchises
to men who are also small businessmen.

Many franchises invest their money because they want to work for
themselves.   Furthermore, they want to affiliate with a going concern
which makes profitable products rather than taking the risk of starting
from scratch by themselves.   These franchise investors are usually looking for a package they can buy for a certain amount.

Franchising is a system of distribution by which you, as a small manufacturer, may be able to widen the market for your products.   You
do this by granting distribution rights or a particular distribution right
to a limited number of *selected* businesses.   In effect, it increases the

number of your outlets that (1) are owned fully by investors other than yourself or your company or (2) are owned in "partnership" arrangements between outside investors and your company.

Franchising offers several advantages to you as a small manufacturer. One is a *rapid expansion*. This method of distribution offers you an immediate source of growth capital without diminishing the ownership of the company. As a franchiser, you are limited only by the standards you set for granting your franchises since investment capital is supplied by the franchisees in most cases.

Reduced management overhead is another advantage. In franchising, the franchisee or his representative is the manager of the outlet. Thus, you do not have to hire managers as you would if you started your own stores. A franchise places ownership and management at the point-of-sale and thereby reduces the parent company's general management overhead.

Another advantage of franchising is that it provides a *built-in community* acceptance of your product. Customers are more receptive to trying a product when they buy it from a local citizen. Finally, franchising provides a *strong success motivation*. Franchisees need to operate efficiently and profitably to protect their cash investment.

Whether your company can enjoy these advantages depends on the type of product you make or the service you offer, as well as the market you are trying to reach, and the buying habits of your ultimate consumer. One or more of the following conditions have been found necessary for a successful franchise:

1. A product or service that is distinctive and has a readily identifiable trademark so that the ultimate consumer will put forth more effort than usual to purchase it.

2. A product that requires installation, an inventory of parts at the local level, and dealer (franchisee) responsibility for maintenance.

3. A new product or service that needs restriction of outlets as a basis for obtaining dealer cooperation.

4. A sufficiently large investment by the franchisee in equipment and/or inventory in return for which there is a relatively adequate markup.

Franchisees need guidance

When the small manufacturer sells a franchise package, his company has to live with it for years. Therefore, it is essential to build a franchise program that will wear well for you and the franchisee. A standard pattern of good service and a unified image with which customers can identify works advantageously for each franchisee and you.   This fact means the franchisee has to have a pattern to go by when he sells your product.   Without guidance on merchandising and profit making even the best program would soon break down into a hodge-podge.

When you insist that your franchisees merchandise in the same way, customers learn to expect the same system of service and the same product whenever they go into one of your outlets.   But even more important to the franchisee is the protection that your system of control should provide him.   When you insist that each outlet run so as to attract customers, you are protecting the investments of all the members in your franchising program as well as your own.

A system of controls should also help the franchisee to operate at a profit.   A system of recordkeeping, for example, helps to spot quickly areas of potential trouble for the franchisee.   In considering the kinds of controls or system you need, two questions can be useful.   First, what is necessary to insure that the franchisees operate according to the plan for attracting customers?   And, what services should you, as the franchiser, furnish to help the franchisee operate at a profit?

Testing with a pilot operation

After you have worked out the system under which your franchisees will operate, the next step is to test it with a pilot operation.   This pilot outlet must be located in a "typical" business area.   *The location for your pilot operation has to be as similar as possible to the kinds of sites your franchises will be using later on.*

Using an average location should give you two things: (1) figures you can use for future planning and (2) assistance in selling the package to franchisees.   When your pilot operation is in an average location, it is fairly easy for your franchise prospects to relate it to their own town or area.

Your pilot operation must feature exactly what you intend to sell potential franchisees.   A person buys a franchise because it is a tested package and because he can see the kind of store in which he plans to invest.

In operating the pilot outlet, accounting procedures and other systems should be maintained exactly as you intend them to operate later on in franchise units.   The pilot operation should run for at least 6 months—sometimes for 12—in order to accumulate figures on sales, cost, and profit.   You use these figures to evaluate the "typical unit" and determine the kind of franchise package you can offer.

How much money

The statistics from your pilot outlet are the basis for determining how much money a franchise unit needs for four things: (1) cash flow, (2) building and equipment, (3) inventory, and (4) profit.

The amount of money that will be needed to insure proper *cash flow* in a franchise outlet hinges on the type of business. If sales build gradually—or if the franchisee has to extend credit to his customers— then he must have enough working cash to tide him over while his store builds a following. Of course, he will need less backup money if the cash return is immediate. Also bearing on his working cash will be the amount of *inventory* he needs to carry in order to serve customers efficiently.

Whether the franchisee needs money for *a building and equipment* depends on the individual circumstances of the franchise. If you pay for the building, for example and lease him equipment, then the burden of having a ready source of investment capital is on your shoulders.

Finally, you will need to use the figures from the pilot operation to determine *profitability*. How will the franchise program contribute to your company's profit? How much profit can the franchisee make on his store? It is vital to keep in mind that unless there is a profitable franchisee, there can be no franchiser.

The franchise agreement

The findings from your operation of the pilot outlet will determine: (1) What the franchise fee will be, (2) the type of franchise agreement, and (3) the amount of working capital, over andabove the franchise fee, that the franchisee will need to get his business started and on a paying basis. Franchise agreements are as varied as the franchise business itself, but agreements usually are of two general types.

One type is the *operating franchise* contract. With it, you allow the franchisee to operate one or several franchises but not to sub-franchise. The second type is the *area franchise* arrangement. With it, you give the franchisee a large territory so he can engage in sub-franchising.

Franchise contracts, regardless of their general type, vary to fit the situation. For example, if the franchise requires a relatively large investment in machinery you manufacture, you may become a "partner" of the franchisee until the machinery is paid for from profits of the business. In other situations, franchisers may provide financial assistance to purchase the building which the franchise business needs to conduct its operation.

In writing a franchise agreement, you will want to work closely with your lawyer.   Your contract should spell out the details, some of which are: the franchise fee and other capital requirements; the area of protection for the franchisee; financial assistance, if any, offered to the franchisee; your control over: (1) location of the business, (2) franchisee purchases, (3) performance, (4) quality and service standards, (5) physical appearance, (6) products carried other than the franchised line, (7) promotional activities, and (8) recommended prices; recordkeeping requirements; management assistance and training; sale of franchise license; termination of franchise.

Understanding the market

In putting together and selling a franchise package, it is helpful to have a knowledge of your market—the kind of person who buys a franchise.   As a rule, he has little or no knowledge of the field he is entering.   He is uncertain about the future and wants to keep his risk as low as possible.   He wants expert attention applied to his investment.   To fill these needs, he buys a franchise.   From it he gets knowledge about a business with which he is usually unfamiliar and volume advantages—purchasing and advertising—offered by a successful company.   Keeping the risk low is important to many franchisees.   They are willing to accept lower returns on their investments in order to minimize the prospect of failure.

It is important to keep in mind that the buyer wants guidance. Furthermore he is apt to feel that he has not gotten his money's worth if the franchiser does not, or cannot, supply the guidance that helps the outlet operate profitably.

The strength of your program depends on how well you help your franchisees.   It is not enough for the franchiser to back off after he has helped, for example, to open the doors to a franchise restaurant.

Selling the package

Three things are important in selling your franchise package—in getting investors who are qualified to work with your company.   They are: the sales brochure, supplemental advertising, and screening prospects.

The Sales Brochure.   Your primary sales tool will be your franchise sales brochure.   Your experience with the pilot franchise unit represents what you have to offer: your brochure tells the reasons why you are offering it. It will introduce prospects to your operation. Because many franchisees have little or no knowledge of your business, the brochure should go into the details. Your brochure can also serve as a public relations tool. It can supply information on your business to trade paper editors, newspapermen, real estate developers, suppliers, and others who need to know about your business. Thus your bro-- chure should be attractive as well as factual and descriptive.

Professional assistance can be helpful in making it so. Keep in mind that a shoddy and cheap-looking brochure cannot sell a franchise bearing a price tag in the thousands of dollars. Professional help can also save you money on printing and mailing by advising on size and weight. In one year, a restaurant franchiser mailed more than 9,500 brochures to interested persons even though it had only 30 franchises to grant.

Supplemental Advertising. In addition to printing a brochure, you will need to advertise the fact that franchises are available. You should select media that are seen by the type of person you want to interest. If you are seeking an unskilled person who can be taught to run a lowinvestment business, then a consistent schedule of ads in a classified section of the newspaper is probably a good approach.

However, if your franchise package has a relatively high cost and requires a specific ability, your advertising should be pinpointed. You will need display ads on financial pages or ads in specialized publications.

Screening Prospects. When your advertising and brochure have generated a list of interested prospects, your next task is to pick the best prospects. You should screen these carefully because the franchiser-franchisee relationship is on a long-term basis, usually 10 years or more. Each applicant has to be evaluated as to character, health, stability, willingness to work, and financial standing. He should be evaluated also on his ability to get along with people because he will be dealing with both the public and employees—and in your name.

Personal interviews are a good way to do the basic screening. Prospects who pass this screening should be interviewed again in a longer and more penetrating meeting. It is a good idea to include the prospect's wife because she plays an important role in the franchise business and should be aware of the problems that can arise in operating a franchise unit.

Your interviews should be give-and-take sessions. Be prepared to answer the prospect's questions about the financial status of your company, future development plans, product quality, price, the franchise territory, payments, type of assistance you will provide, advertising costs, and lease agreements.   Sometimes, the prospect may bring his lawyer to the early interviews.   If he does not, you should insist that the franchisee have a lawyer to advise and represent him when the contract is signed.   Such advice helps to prevent misunderstandings later on.

Training the franchisee

After the agreement has been signed, your new franchisee will be eager to get started, and he should be moved into his training while his interest is still high.   Training him in your way of doing business is important because you and your other franchisees suffer when one franchisee falls down on the job.   Training should stress that the rewards are there for persons with ability, initiative, and endurance.

You will want to keep in mind that the individual franchisee's actions, in the long run, determine the success or failure of a franchise program. When he reflects the proper training in the management of his outlet, franchising can indeed be a lever for accelerating the growth of the franchiser.

X. The Story Of Associated Ideas

The story of how Associated Ideas began is not a dramatic one. There were no blinding flashes of insight that inspired it, no taut meetings of grim-faced men searching desperately for a solution to a sudden emergency. Rather, it grew out of a lack—a bone dry, frustrating inability to be an expert in every field. Ideas were plentiful and the desire was strong but still years of work on potentially fruitful inventions proved disappointing at one stage of another in the operation between inventing and marketing. This, in spite of the fact that significant amounts of money and a large amount of time were spent looking for a workable system to bring an invention idea to full production.

But there is no simple solution. The talents and experience needed to make a pratical business out of inventing can cover, at one time or another, half-a-dozen extremely complex fields. A man who understood this only too well was co-author Terry Fenner. As a research chemist for Southern Regional Agricultural Laboratory, he knew the advantages of the group approach to problem solving—but also the disadvantages as many men worked separately on personal projects that if successful could mean promotion or a raise. Conditions were friendly and cooperation maintained but because new ideas could affect salary, there was a lack of trust needed to bring out the best efforts of the workers.

Self-taught in many technical areas, Fenner was not afraid to tackle the job of learning the patent business by himself. He had worked on inventions that were patented by the laboratory under his name and also had his own projects patented. So why not carry forward his ideas alone? He knew the areas of competence that had to be mastered and set about learning them - how to write patent applications, marketing outlets, court proceedings, royalty rights, etc. And it took years. He finally realized one man could not learn the whole field alone.   Too many mistakes were being made and useful ideas being wasted because patent applications were weakly written, or the wrong companies were contacted for marketing, or he ran out of money.

From experience in the research laboratory, he knew the efficient results gained through the group approach—experts in many fields contributing their experience to work out a complex, multifaceted problem. The missing ingredient for true productive effort seemed to be harmony of personalities and trust among co-workers.   If he could organiize a group along these lines, he felt it would be an ideal situation for creative effort.

Finding Interested Persons is Easy

Out of all the capable men at the plant, Fenner approached one, Hozenthal, whom he knew would be harmonious in personal relations —active in ideas but not overbearing.   As a mechanical engineer, Hozenthal would add another dimension to the group's portfolio.

Finding other interested persons was relatively easy.   Fenner had already worked with private industry in developing and testing new products at the laboratory.   In addition, three relatives were also able to contribute their talents as a chemist, businessman and professional artist.   The problem was more one of selection than searching. Fenner felt a small group would be more manageable.   He was fortunate in his choice as proved out later (almost 70% of the original group remained intact after 7 years and others who came in and dropped for one reason or another are still open for consultation on specific projects).

There was quite a bit of mutual congratulations among members at the group's first meeting.   Enthusiasm ran high, almost rampant as they set out to solve any and every problem that came their way. But they were not really sure what constituted a "fruitful problem" and so sent out queries to industry and business offering their services in research and development.   Some members had a few pet projects to be worked on, but group decided to use the scatter-gun approach, and out of it something worthwhile should come up.   Any and all suggestions for projects were welcomed—Associated Ideas was ready to tackle anything from discussing the pros and cons of perpetual motion machine to putting ketchup in an aerosol can!

First, their very enthusiasm was an obstacle to concentrated, effective effort.   So many proposals were considered, they could not possibly handle them all, and met later on with the inevitable let down and discouragement. Unfortunately, the group was comprised mostly of "idea men" and did not have the balancing factor of hardheaded businessmen to give a sometimes needed pragmatic approach. They relied upon personal opinions of the members, not upon facts presented by an experienced marketing agent or economist. For instance, not knowing the techniques of the highly competitive, stylized toy market, they spent considerable time in trying to break in this field and met with little success. It was not until years after that A.I. found out how much of a "closed system" is involved in toy manufacturing.

Importance of Setting up Workable Goals

Another early mistake they made was in not setting up workable goals. Though everyone was contributing their talents part-time and meeting once a month, they thought they would be able to offer their services as a research and development to small and medium sized businesses that could not afford one of their own. With no full-time member to make contacts and coordinate activities, this soon proved impossible. However, out of it did come several jobs as consultants to answer specific problems in brainstorming sessions. It helped establish their name in the community and through contacts with different businessmen many of the members were able to get better jobs or pay raises as participation in such an active group added to the person's dossier.

The main stumbling block of the group in the beginning was a too loose-knit organization caused, no doubt, by a lack of any previous experience. Its primary aim was not just to make money, but to exchange information about the various facets of inventing, to offer a community service for fledging inventors, to enjoy the sociability of each other's company in a common interest, and eventually, to begin the serious business of promoting and perfecting a few selective, worthwhile patents.

With no definite guidelines to follow, it took almost four years before the group finally decided to restrict its intellectual pursuits to only those that would be followed up with patent applications. This, because of the expense involved, served to wean out the more frivolous projects, and for the first time, an efficient return of time and money was realized.

Though the group criticized themselves on many occasions for giving too freely of their time and talents, in retrospect their contribution to the community had far reaching effects. Some were simple things, like helping the man who had been working on an ant trap for 20 years, and even had three patents on his device to protect it from "unscrupulous businessmen." But what could he do with his lifetime dream, when the technology of insecticides made his invention obsolete? Rather than turn the man down with the obvious argument that there was no longer any market for his ant trap, A.I. made an ingenious ant circus out of the device with just a few modifications. They turned it into an interesting toy complete with trapezes, slides and trap doors, that could provide hours of amusement for youngsters.

Contribution to the Community

Of major importance was the promoting and organization of a research and development institute for the State of Louisiana. Literally hundreds of people were contacted to form a non-profit organization that would aid business in technical research and investigate new uses for natural resources in the state. But just as it was about to become a reality another group with state backing also saw the need for this kind of institute, and as they could command the personnel and facilities needed for such a large scale operation Associated Ideas threw its resources, contacts, and good will behind the newer organization to help get it off the ground. It is now a $5 million institute that is growing with almost unlimited potential.

Associated Ideas' TV Program

Involvement in community affairs began early in the club's history. A fairly lengthy story in the local paper brought many inquiries for help to the club, as well as new members. But, perhaps, the most effective agent for advertising was a once-a-week 5 minute program on TV. Though this was scheduled for in the morning on a show that was predominantly oriented to women, it produced queries from both sexes long after the program was finished. In fact, one of the announcers on the show said it had as much continued audience interest as any series they had carried.

However, getting enough interesting lively material was a continued headache for the members who had to stop work on some of their personal projects to prepare TV material. Demonstrations were made of both practical inventions (such as flame-proofing cotton, ultrasonic generators for cleaning clothes, disposable raincoats, etc.) as well as "Rube Goldberg" devices illustrating what not to do (i.e. a motorized back scratcher that went haywire and a trick leash to keep a dog from running away from its owner by raising its back feet).

Even though the show was amateurish as far as acting talents of the members—on the first show a heavy barbecue grill was dropped on the announcer's foot in full camera view—the success of the program illustrated the wide interest of the public in this field. The free advertising was invaluable and though shows were time consuming, the name of Associated Ideas was well implanted in the city and acted as a background introduction for the club when it later approached business and civic officials for promotion of the research institute.

Because of the avalanche of calls for information and help that resulted from the TV program, A.I. set up definite rules to conserve its time and efforts and aid in being more selective in accepting consulting work.

First, A.I. would interview the interested party with no charge. Then a decision was made whether or not this was the type of problem it could solve and which member should handle it. If the problem was one that involved the whole group, then an extra charge was made. Cost was determined upon time and number of members engaged in it ($2.50 per hour per person). In this way, only persons who were seriously interested in perfecting their idea would come back for continued service. Furthermore, if the group as a whole thought the idea was trivial, they would vote not to accept it.

From consulting work, original membership fees of $25 apiece, and monthly dues of $1, A.I. had barely enough to cover operating expenses of searching, submitting patents, stocking library, correspondence. Finances were limited but money did not seem to be a problem source for three or four years. It was not until a lawyer asked to buy membership in the group as a future investment that they realized the whole organization could erupt if one patent suddenly hit it big and members began asking for their proper share of the money. The investment potential of the group lay in the fact that if A.I. paid expenses to submit and perfect a patent, then the member whose name the patent was in would assign 50 percent of royalty rights to A.I. This is not so important in the beginning as there is a certain time lag between submission of patent application and contacting of companies before the patent begins to pay off. But, sooner or later, a decision about division of royalty rights will have to be made.

Incorporation

Associated Ideas, on the advice of the lawyer, formed a corporation with a limited stock value of $1,000 with shares to be sold equally to all active members.   Stipulation was that members had to bring their dues up to date, pay the additional fee for incorporation $100 apiece for which each received ten shares of stock) within one year's time, and materially aid in promoting the organization by attending meetings on a regular basis and offering services that were asked for.

If one were to assess the goals of Associated Ideas in the first five years of operation, it could be said that about 90 percent of them were realized.   Of course, as an open-ended operation, its work is never completed, but the basic structure has been developed and proved sound, and provides a foundation for future expansion.   Perhaps the greatest weakness the group had to overcome as had been mentioned before, was the over-emphasis on exploring potentials of new ideas for the sheer interest of it, without enough practical application to make it pay off.   For too long, the members attacked projects more as a hobby than a well balanced business effort, which kept enthusiasm high, but must be rated as a minus in reporting the progress the club made towards its original goals.   The following chart is the rating the group gives itself upon its own examination:

An Evaluation of Obtaining the Goals Proposed When Associated Ideas Was Established

Goal	Successful	Unsuccessful
Mutual help to Its members	In addition to becoming more familiar with the invention field, 3 members each received I patent and are marketing their inventions, 4 members received job promotions because of help received	
To try the group approach to the field of inventions	Associated Ideas now has 4 active projects underway in addition to writing this book, plus I issued patent and 5 patent applications	
To make money in the field of inventions		Associated Ideas received consulting fees on various outside projects but has not made a great deal of money in inventions as yet
To educate its members	Associated Ideas has sponsored lectures, movies, and demonstrations in fields of interest to the inventor. One member through the experience gained has taken the examination to practice as a patent Agent	
To try cross fertilization of ideas from various fields on inventions	By having people in many fields view an invention in the eyes of their own profession after the invention is gretly improved	
To make inventioning interesting and enjoyable	Eighty percent of our members have been in the organization over five years	

In a more detailed analysis, the next list shows the range of projects undertaken. Though only partial, it serves as a criterion to show in specific examples the reason why the members feel they have been adequately rewarded for their group efforts.

Projects Attempted by Associated Ideas (Partial List)

Pending	Successful	Unsuccessful
"Handbook for Inventors"	A five minute television program, once a week, interviewing local inventors WDSU-TV (well-received) ran one season.	"Fascination Board" Associated Ideas attempted to sell the plans for a board to be built at home to amuse young children. A plastic unit already assembled "Busy Box" later came out and sold very well.
Merchandising a sequin applicator for applying sequins to a glued surface	Associated Ideas brainstormed or invented an improved pile driver pad (consulting fee obtained)	Duplication of American products by an importer (he would not sign a contract to pay)
Promotion of our patented "cutting mechanism"	An information and education program to inform inventors and the community of the role of inventors. Valuable industrial contacts were established	Coffee pak—placing coffee grounds in a "tea bag". Better tasting instant coffee killed this product before it was marketed
Merchandising a disposable plastic raincoat	Aided in the establishment of a $4 million research institute that will aid our nation's inventors.	
Promotion of an improved fuel cell system (patent pending)	Obtained one patent, filed five others	
Promotion of methods to manufacture aluminum more economically (patent pending)		

Questions and Answers on the Advantages of the Group Approach

After so much ballyhoo about the advantages of the group approach, it would be good to describe the disadvantages, or at least limitations, to the pooling of common talents and time.   These problem areas are fairly common to all small groups, and while they do not cover every area of potential pitfalls, they definitely represent questions that have to be resolved.   After describing the problems that may arise, an answer will be given that Associated Ideas has worked out, based on years of practical experience.   Though other answers are certainly possible, these reflect the groups personal experiences.

I.   Part of your time, effort, money is given up to the potential profit of another.   What is there that you can get out of it that would make you bother with the effort when, perhaps, you could make more efficient use of your time by following your own ends.

Ia.   First, there is that intangible but vital ingredient called "motivation."   Often, when discouraged or stuck on a problem, working it out or talking it out with another person who is knowledgeable will help make the answer clearer.   Also, the creative process is enhanced by competition, and fairminded criticism can be an aid to keep you from going too far off the track.

Of more practical importance is the sharing of techniques, resources, library (magazines, periodicals, books), and the pooling of money to increase purchasing power.   Group affiliation also gives better status and recognition when seeking financial backing from businessmen in the comunity.

To make profits more equitable, shares could be given to those who did the most work on a project.   Proportions could be arranged by corporate stock or percentage of profits from the invention when marketed.   This must be agreed upon in the beginning, however, and set down in writing and witnessed so there will be no hard feelings afterwards.   Some flexibility should be allowed to keep harmonious relations because a too "cut and dry, businesslike attitude" could create friction or distrust.

2.   Along this same line, there is a definite time lag between organizing the group and receiving royalties on the first invention. It may be many years if the group is starting off from scratch with no previous experience.

2a.   A time lag is inevitable, even if working by yourself—there is no way to get patents through the Patent Office any quicker.   However, after one invention is started, another is picked up so that there are always things in the mill.   To keep interest up, it is better to have at least four inventions in some stage of completion, but hold the group down because it gets unwiedly and costly to cover too many patent applications at once.

While patents are processing, the group could be engaged in consulting work or research through the Small Business Administration or by bids for government contracts.   Every large corporation has a small business representative and government encourages sub-contracting to lesser companies.   In some instances, the group may have to be incorporated, but if it can prove its ability to do the job satisfactorily, arrangements can be made, even on a part-time basis.   Once again, the persons actually doing the work should be compensated accordingly, with some profit going to the club's coffers for future funding.

The social aspects of the group are also to be considered.   If there is a harmonious relationship and common interests, many meetings will be held just to enjoy each other's company and not always be "putting their noses to the grindstone."   One reason why Associated Ideas stayed on the social plane for a long time is that no one wanted to go full-time to get outside work and thus speed up the growth process.   This is a decision each individual chapter will have to make depending upon the choice of its members.

3.   When a project is given over to the group to discuss or carry forward, there might be too much of a "let George do it" attitude.

3a.   This can, and does, happen in almost every organization; one or more persons seem to take on the burden of responsibility far in excess of the rest.   One member of AI worked so hard, even going without sleep and neglecting his family, that the group hated to give him an assignment because he got carried away with his own enthusiasm.   But it is usually the other way around—where the problem lies in getting the work done on time.   A practical way to handle this is to bring up delegated assignments at the reading of minutes for each club meeting.   Then ask the people involved what they have accomplished so far, what their problems are, do they need more help or time, etc.   This reminder serves as a social pressure that people will usually respond to.

There is a certain drawback to "talking out" a creative idea.   Some interior tension has to be maintained to dredge up the most original ideas from the subconscious. There is a time for brainstorming and a time for keeping quitet and searching your own mind for the best answer. The talking business can be carried too far, and then an idea is "committed to death."

4. How do you keep nuts and time-wasters out of the group?

4a. Sometimes you don't. Like it or not, some effort should be given back to the community. Not all inventors that come to you for help will be the most practical or sensible. But it is still the club's duty to guide them as best it can within reasonable limits of time and effort.

As for personalities within the club who may be just riding along for social events or prestige of membership, make dues, collections and attendance at meetings necessary. That will separate the genuinely interested workers from the hangers-on. Further, it should be understood at the beginning that to keep the group harmonious, it may be necessary to drop some members or change the makeup of the group to give it better balance.

5. Why all the bother about electing officers, following established formats, collecting dues, etc? Why not just form a congenial group of co-workers and engage in common activities as the desire or opportunity arises, on or off the job.

5a. That would be the best starting point for forming a group. But some important things have to be kept in mind. First, will the members remain harmonious and unselfish, or will some take advantage of the others realizing that on the job new ideas mean promotion or raises. This is the way a research laboratory team is supposed to operate, but most often the team is limited in creativity by the regime of the place in which they work, or by lack of trust among workers or between management and personnel.

Also, if a checking account is to be kept to finance ventures, banks or other lending institutions, require that some one be appointed responsible for check signing or indebtedness. Loans, either from private captital or the Small Business Administration, also demand that books be kept to show accountability. The arrangement of officers is best understood as a necessary division of labor that is commonly accepted and expected in most business transactions.

6. Knowing there have been other organizations established, both on local and national levels, why would not they be just as good as or better than Associated Ideas?

6a. Without wishing to demean their intentions, or even effectiveness, the main disadvantage would be that they do not offer a full scope of services to the inventor.   There may very well be some on a local level that are similar in nature to Associated Ideas, but in five years of searching, the authors have not been able to find a going, national organization that satisfactorily answers all the needs of independent inventors.   If there were, A.I. members would have joined a long time ago and saved themselves literally hundreds of hours of work.

Here are some of the drawbacks that have been encountered in investigating opportunities of other organizations: One national company offers reviews of inventions for a fee, but comments are mostly favorable or too general to give any in-depth analysis of potential flaws; also, their involved method of recording date of disclosure can be done just as easily and effectively with one witness.

Local organizations that will help to market an invention for a fee seem to be profit-making, not profit sharing, so some of the money invested may go to the head of the organization, not to the members and they not be used for the fullest benefit of the individual inventor.

Other groups are specialized and want inventions only after a certain stage when profits are more reasonably certain, and so do not encourage creativity from the very beginning or the idea.

Lastly, because individual inventors do not have a chance to become officers in the parent organization, full interests and capabilities of all members may not be realized and it becomes more a business arrangement than a non-profit society for the good of the whole field.

Associated Ideas International

Many other persons besides those in A.I. have realized the need for a full service national inventor's association.   Requests for membership or affiliation with A.I. have been received not only from all parts of the United States, but other countries as well, including Mexico, Switzerland, South America, Canada, etc., but until now it was considered just not possible to expand A.I. on a nationwide scale, much less worldwide.   The costs for publicity, organization, personal meetings around the country would be prohibitive.   But gradually as more names kept coming in, as A.I. itself grew more productive, a breakthrough was gradually made and a concerted plan of action became possible. Thus, this book was written to tell the story of the individual inventor and Associated Ideas International (A.I.I.) and how the two could complement each other. An integral part of this was the methodical collecting and listing of inventors, researchers, market outlets, businesses, trade organizations, etc. that would cooperate in the venture to open communication lines in the many different disciplines to which an inventor must have access. With over 200,000 independent inventors in the United States alone, the market is surely open, and indeed, should be self perpetuating.

The rules for becoming a member of Associated Ideas International are brief, and the benefits are many. By accepting an invitation to join, you will be part of an association devoted completely to inventions and mutual help for inventors. You will join others throughout the world who believe that a group approach to the field of inventing is the one practical method to insure success, especially since patent laws are becoming stricter and the technological information explosion gives all indications of expanding on an even greater scale in the future.

Local chapter activities will provide first-hand knowledge of individual inventor's needs and progress and enable each other to meet and exchange ideas. In addition, you will be invited to attend regional national and international A.I.I. meetings and seminars including the bi-annual convention that will feature inventor's exhibitions and demonstrations of experimental research likely to affect the future of us all.

Principal cities of this country would each have a chapter composed of up to 16 members divided into work commitees that had interests common to the people engaged in a project. Depending upon the number of interested parties in a specific locale, more than one chapter could be established in a city. The international organization would reflect the views of the majority of the members by having a president and geographic representatives elected from the general membership. In addition, a full-time staff to handle and promote the best interests of the membership would be employed.

Requirements for Membership

The only requirement imposed on individuals by A.I.I. is that a member shall have a genuine interest in inventions. Local chapters will be free to impose any other restrictions and other limitations that best suit their particular situation. Thus any member in good standing in a local chapter shall be given the title of senior member with all rights and privileges. But others can also join A.I.I. as associate members with full privileges. This may happen when for good and just reason an inventor does not or can not join a local chapter.   For instance, retired persons having wide experience in the business world would be extremely valuable to A.I.I. or local groups, but because of their age or other reasons they may not want to engage actively in a small group, but would like to contribute their talents on a less strenuous basis.   Their talents would certainly be an asset and they should be highly encouraged to join in a manner that would not put undue pressure on them.

High School and College Chapters

Because many people often exhibit inventiveness at an early age, and to encourage youth to enter the field as either a vocation or avocation, chapters would be also situated in high schools and colleges with attractive student membership rates of $2 per year and with students still eligible for all the privileges of full membership.   This could work in hand and hand with such organizations as Junior Achievement Clubs that are sponsored by businesses to create, produce, and market products.   Through A.I.I. they could learn to develop much more sophisticated products, or even serve as distribution outlets for inventions already patented that need testing or wide public distribution.

Person-to-person contacts with manufacturers throughout the world will also be possible as chapters are founded in principal industrial cities.   For example, an inventor in New Orleans would ask a member of the chapter in Detroit to present an improved car engine design to an auto company there.   The Detroit representative would be reimbursed for his time and expenses for providing a service the New Orleans member could not have afforded without a much greater outlay of cash.

Other chapters may want to take on specific duties that they have a unique ability to handle.   One could act as a marketing "sounding board" for the entire organization.   Another chapter, whose interest lies in publicity, could perform the function of publishing the organization's magazine, or serve as experts in such highly stylized fields as toy design, electronics, space, ocean technology, plant patenting, etc. Of course, each chapter would be paid for its services.

Where possible A.I.I. will investigate and evaluate private sales agencies and brokers and recommend those who are qualified.   Some of these appear in the Appendix, such as the Battelle Development

Corporation and Arthur D. Little, Inc.   The list of manufacturers who want inventions in specific fields will also be continuously updated to show what fields offer the most promise of success.

By necessity A.I.I. will seek the betterment of the inventor by lobbying for government legislation, promotion workshops and training seminars, developing more effective community relations, backing research and development programs and speaking up for the inventor where his concern is affected by public issues.

Advantages of Membership

One of the features of A.I.I. will be a magazine covering the latest research and needed inventions in the various scientific fields plus other topics of interest to inventors.   Additional advantages of All membership includes:

a.   A chance to have your invention proposed for consideration in a contest with the most meritorious invention being patented at no cost to the chapter.

b.   Participation in an international convention, a bi-yearly convention, and sectional meetings to hear featured speakers on latest technology demonstrations of inventions, seminars to meet and exchange ideas, contacts with manufacturers, representatives, etc.

c.   Special discounts on items of interest to inventors will be offered that may be worth many times the yearly international dues.

d.   Underwriting of money borrowed by local chapters for research and manufacturing of inventions considered outstanding by the organization.

e.   Listing of member's names in the Associated Ideas International Directory of Inventors.

f.   Special services such as job placement and invention review and evaluation.

How Much?

How much would all of this cost?   Dues for membership in A.I.I. would be $10 yearly (students $2).   Local chapters will assess their own fees to cover expenses of meetings, projects, patent applications, stenography, typing, postage, miscellaneous expenses. (Most adult chapters vote approximately $15 per person per annum for the first year for expenses.)   Successful chapters could later incorporate under the laws of their own state and issue stock against inventions developed by themselves.

An A.I.I. inventor would not have to relinquish any title to his invention, however, if he wished his local chapter to undertake development, a suitable percentage would have to be assigned to the chapter, usually 25 percent.

How to form a chapter of Associated Ideas International:

A petition for establishment of a chapter from a minimum of five members shall be mailed to International Headquarters, 4447 Lafaye Street, New Orleans, Louisiana 70122, United States of America. No prescribed form is required or necessary. A recommended maximum number for a chapter's membership shall be sixteen.

Elections are to be held and the elected officer's names and addresses —president, vice-president, secretary, and treasurer—are to be forwarded to International Headquarters. The name of the chapter will also be forwarded.

The annual international membership dues of $10 (ten dollars), students $2 (two dollars), shall accompany individual and chapter petitions for membership. The name and address of the student's school should also be included in the student's application.

In the case of individual membership applications, the names and addresses of individual persons residing in the same geographic location received as a result of publicity from the book shall be collected and exchanged so that local arrangements can be made for an organizational meeting and election of officers.

It is A.I.I.'s aim to establish a unique Technical-Coordination-Information Center that would be of invaluable benefit to the whole economy. For the first time, the composite talents of thousands of inventors would be utilized efficiently.

Additional information may be obtained by writing International Headquarters at the above address.

How to Organize An A.I.I. Chapter First Meeting

Pre-meeting orientation

Select a group of people you think would be interested in inventing. Write them and explain about Associated Ideas International and how it could benefit them. Set a date, time and place for an organizational meeting (possibly your home).

The first meeting's agenda

Organization

1.   Nominate and vote a chairman for the meeting, who then takes charge.
2.   Discuss how your chapter will operate.
3.   Establish ground rules.
4.   Discuss and select an invention, product or service your chapter would perform.
5.   Assign tasks to bring this into fruition by the next meeting, if possible.
6.   Establish a research and development committee for the selection of future products.
7.   Select your chapter's name.
8.   Apply for your charter's membership in A.I.I.
9.   Nominate and vote a slate of officers.
10. Discuss your chapter's need for capital and how you will finance your projects.
11. Review.
12. Adjourn

Review

During this first meeting you have accomplished a great deal.   Let us review what has happened so far.

You have met your fellow members.

You have learned what A.I.I. is all about.

You have established ground rules.

You have decided on the invention, product or service you will sell.

You have learned how a corporation is formed.

You have decided on the name for your chapter.

You have applied for a Charter and signed the application as a Charter Member.

You have set your capitalization goal.

You have learned the responsibilities of serving as a member of a worldwide inventor's organization.

Get some publicity about the formation of your chapter and the election of officers in your local press.   Newspapers are always anxious to have "new" news to print so don't put off contacting the city editor too long.

Possible Agenda For Second And Subsequent Meetings

Call to order

President

A. Secretary records attendance.
B. Secretary reads Minutes of last meeting. Vote approval.

Reports

A. President—General condition of the chapter.
B. Treasurer—Profit and Loss Statement, Balance Sheet.
C. Vice President—Sales and Production.

Old Business

Subjects presented to the members previously on which action has not yet been completed. The individual concerned describes current status of the subject and makes a recommendation for action.

New Business

A. Subjects not previously considered that require action by the members. The President asks for any such subjects as may have come up since the last meeting.
B. If study is required, he refers the subject to the responsible person for report at the next meeting. If action can be taken now, he asks for an appropriate motion.

Adjourn

A change in meeting style—for example, a movie or demonstration on a latest research advance by a large company—is suggested to keep members' interests up.

An entire chapter may meet once a month and a work committee on a project as often during a month as necessary.   Use the telephone for committee work rather than calling frequent meetings.

ASSOCIATED IDEAS INTERNATIONAL

Application for Charter or Individual Membership

TO: International Headquarters
Associated Ideas International
4447 Lafaye Street
New Orleans, Louisiana
United States of America

Gentlemen:

On this................................the................of........................, 19........
      DAY OF WEEK   DATE   MONTH   YEAR

FOR STAFF USE ONLY

APPROVED:

By:...
           EXECUTIVE DIRECTOR

Issued Charter No....................................

Date...

We, the undersigned charter members, do hereby petition for a charter to operate under the By-Laws and official policies of Associated Ideas International, a chapter to be known as:

....

PRINT NAME EXACTLY AS YOU WISH IT ON CHARTER

....

STREET ADDRESS OF CHAPTER   CITY   STATE

and enclose our respective annual membership dues of $10.00 per member (students $2.00). Dues may be pro rated on a monthly basis beginning January 1.

Our Elected Officers are:

President ....

Vice-President ....

Secretary ....

Treasurer ....

Individual or Charter Members must sign on reverse side.

XI. 100 Needed Inventions

Inventors often ask "What inventions are needed so that I can begin at once to think of practical solutions." One need only look about in his own job, home, or recreation to find needed inventions. Trade associations are often a source of good advice on such inventions.

Inventing significant improvements in existing products is probably the most lucrative field for the inventor. Notice the word "significant." Disadvantages or annoyances present in existing inventions which, if eliminated, would greatly increase sales or influence potential buyers are significant improvements. (Example: The electric starter for cars vs. hand cranking).

The following are 100 needed inventions submitted for your consideration. Before starting work on these or any invention, write companies in the field to verify that the need still exists. Often times, suggestions will be made by these companies on what has been done in the past and the most promising approaches towards solution.

1. An automatic home clothes folding machine.
2. The American Humane Association offers $10,000 for a human trap for wild animals. The reward will be paid to the inventor of a trap adaptable for use for all animals, or the prize will be divided evenly between the inventors of traps for small or large animals, as it may not be possible to develop a trap for both mink to bear.
3. An inexpensive device to produce a three dimensional effect for a television set.
4. An inexpensive means for projecting a television picture from a set onto a wall for lifesize television.
5. An inexpensive method for "piping" a television program onto several screens in different parts of the house from a single TV set.

6. A method to produce air conditioned and/or heated clothing.
7. A tractor designed positively not to tip over.
8. An improved method of repelling birds from airports and public places.
9. An improved electric trolley bus connection to overhead wires, overcoming problems of disconnection, going around an obstacles in the middle of the street, sleet formation on the wires in winter.
10. A process to overcome the terrific heat–corrosion problem and improve the reaction efficiency of magnetohydrodynamic electrical generators.
11. An inexpensive workable fuel cell for the home.
12. A method for changing carbohydrates and fattening foods into a non fat-producing form.
13. A method for changing indigestable carbohydrates into digestible foodstuffs to help feed the millions of people in the world.
14. An improved method for detecting submarines.
15. An improved method of communications between friendly submarines.
16. A synthetic, non-refrigerated, ice skating rink.
17. Plastics with an extra high strength-to-weight ratio, or molding procedures to produce same.
18. An inexpensive process to manufacture Polaroid glass.
19. A chemical in gasoline that would react with and eliminate the smog-producing products of combustion.
20. A radical new design for service stations for a crowded area where property costs are very high.
21. A method of drilling for oil so that the bit would not need to be raised and replaced so often.
22. A better method of making oil wells safe from explosion and fire.
23. A cheaper and more permanent method to prevent termites from digesting wood.
24. A method to recover more of the actual value of coal, gas, etc., possibly based on a colorimeter design of furnaces, boilers or stoves.
25. An improved detector for locating leaks in high voltage transmission lines.
26. An efficient low-to-high voltage conversion and power distribution system for extra high voltage electrical power lines.

27. A cloth-fastening process and method of producing clothes that will eliminate much of the sewing and hand operations now necessary, or a radically new method to produce clothes.

28. An economical process for producing alumina from soil containing a low grade alumina ore.

29. Cheaper processes to produce aluminum.

30. A longer lasting electrolytic furnace.

31. A method to cut down on the amount of cryolite that is vaporized during the aluminum electrolytic reduction process.

32. A method to reduce the amount of electricity needed to produce aluminum.

33. A method to minimize airplane noise level without affecting the power of the motor in any way.

34. An improved airplane needing less runway in take off and landing.

35. A whitening agent for teeth, possibly based on a dye or optical brightening agent.

36. A substitute for silver nitrate that can be used all over the mouth particularly the front teeth to check tooth decay.

37. A finishing method or process to make cotton textiles permanent press and wrinkle resistant without impairing strength.

38. A white replacement for carbon black in rubber.

39. A signal that would inform a person that someone was trying to reach his number while the phone was being used.

40. An improved process and inexpensive equipment for converting grasses and legumes into a cured, dry feed for cattle without all the complications the present systems have.

41. Monitoring equipment to detect malfunctioning on grain drills.

42. Automatic flow control in harvesting crops tied in with ground speed of the equipment and crop density.

43. A radically new fertilizer production process that can be produced without huge costly plant installation.

44. A radically new process to heat and cool at half the cost of present day systems.

45. An inexpensive substitute for silver in making mirrors.

46. A non-destructive service life test for electrical components and tubes.

47. A high temperature metal substitute for platinum having platinum's electrical non-arcing and inertness properties.

48. A chemical or enzyme to aid in the digestion of wood into a

soluble cellulose fraction, or a more easily produced cellulose paper pulp.

49. An inexpensive solvent for cellulose.
50. A method to hydrolyze wood into sucrose and other edible sugars.
51. An inexpensive process to produce metal whiskers of exceptional strength to be incorporated into molten metal to improve the strength of castings.
52. A process to produce a cast iron that would bend under stress rather than break, possibly by mechanical homogenization of the molten iron.
53. Metals with greater strength and lighter weight via improved metal structures or casting techniques.
54. An economical process for producing iron from a low grade iron ore.
55. A method for preventing train and trolley wires from breaking in cold weather.
56. An alloy of copper or a mechanical process to make copper as hard as steel yet still retain its beautiful color.
57. An inexpensive 165° panoramic-view camera.
58. A camera that will produce a perfect picture, possibly at the expense of a little wasted film.
59. A method for converting speech directly into typewritten characters in a readable manner and vice versa.
60. A tobacco-treating process to yield a tobacco that produces no tar or harmful lung irritants and leaves no ash.
61. A safety device for manufacturing plants that have the danger of fine particles or solvent explosions.
62. A permanent integral waterproofing agent for gypsum in sheet rock that will not affect the strength in any way.
63. A method to increase the flow and delivery rate of pipeline material.
64. Improvements overcoming the problems of using a public toilet (raising or lowering the toilet seat, touching the flush toilet handle, the sanitation of the seat or bowl).
65. A less expensive and possibly better method of binding books.
66. A breakproof chinaware for everyday home and restaurant use, possibly based on a hetrogeneous nucleation post treatment, like pyroceram or on an unbreakable glass process.
67. A radically new design for an automobile battery so that it will

last the life of the car.

68. A method of hardening steel dies that will not alter dimensions.

69. A special drill for aluminum metal that will not clog up and produce oversized holes.

70. A tool that could be used to extract broken drills and taps from castings.

71. A better method of ventilating goggles.

72. Uses for the byproducts such as pulp from grapes used in making wine, coffee grounds from instant coffee, etc. Check your local manufacturing plants for specific byproducts.

73. New uses for dry ice or carbon dioxide and effective methods for controlling the properties of dry ice.

74. A method of neutralizing static electricity, particularly in cold dry weather, which causes problems with feeding paper to printing presses.

75. A method to prevent glass mirrors from fogging up during rain or moist conditions.

76. New or improved super-performance textiles (fibers, papers and plastics).

77. Improved sources for fixed power installations (magnetohydrodynamic, thermoelectric and thermionic, etc.).

78. Improved power sources for transportation (storage battery, fuel cell, improved engines, etc.).

79. New methods of water transportation.

80. Effective appetite and weight control aids.

81. Improved uses of the ocean for mining, extraction of minerals and farming, etc.

82. Automated housekeeping and home maintenance.

83. Improved methods of teaching adults to read.

84. New kinds of very cheap, convenient and more important reliable birth control techniques.

85. New and improved materials for buildings and interiors.

86. Radical improvements in earth moving and construction equipment.

87. Improved learning techniques for acquiring skills.

88. Radically new methods of communications.

89. Inexpensive (less than 1¢ a copy) black and white and color reproduction systems for home and office.

90. New permanent sources of power for individual items such as lights, appliances and machines.

91. Inexpensive worldwide transportation of humans or cargo.
92. Improved and less expensive road building techniques and materials.
93. Direct control of individual thought processes and dreams.
94. New techniques for preserving food.
95. An auxiliary power source for aircraft lasting a few minutes until a plane can be safely landed.
96. An inexpensive substitute for water in industrial processes.
97. An improved anti-friction material for bearings.
98. An ingredient for concrete to overcome cracking or a surface treatment to accomplish this result.
99. An automatic carriage-return attachment for manual typewriters.
100. Important improvements in existing products to make them more valuable and salable.

To tackle these and any invention problem, find out all there is to know on the subject. Check with your library, call on or write companies in the field and see if the invention is still needed. Get the whole picture. Remember: "Chance favors the prepared mind."

Appendix

Disclaimer of Responsibility

Associated Ideas, Inc., the authors, and the publisher take no responsibility whatever as to the financial stability, integrity or any eventuality that might arise between the inventor and any individual or firm listed in this book.

Companies Willing to Consider Inventions

The following companies * will be happy to consider, for possible purchase, worthwhile inventions submitted to them. The inventor should first write for a disclosure form before submitting his invention or idea, unless otherwise stated.

I. TOYMASTER PRODUCTS CO.
141 Lanza Avenue
Garfield, New Jersey
Mr. Samuel Beder, *President*

Manufacturer of corrugated fiberboard toys, games, storage boxes, chests, wardrobes and furniture. Novelty items. Christmas fireplaces, Christmas card-holders, ornament boxes.
Distribution: Major catalogs, department and chain stores.
Interested in all items made of corrugated fiberboard or that can be constructed of wood plus corrugated fiberboard.

2. WILLIAM PRYM, INC.
Dayville, Connecticut 06241
Mr. George W. Salzer, *Marketing Manager*

Light wire and strip metal products selling primarily in the home sewing industry.
The inventor should write for a disclosure form if the invention does not have a patent applied for.

3. PENTAPCO, INC.
963 Newark Avenue
Elizabeth, New Jersey 07207
Mr. Bruce L. Kline, *Vice President*

Manufacturers and packagers of all types of sewing and household smallwares.

4. H. FISHLOVE & COMPANY
712 North Franklin Street
Chicago, Illinois 60610
Mr. Irving H. Fishlove

* At the end of this section is an alphabetically arranged index of the products manufactured by the above-mentioned companies and areas in which they are interested.

Jokes, tricks, gags, magnetic novelties—anything unusual. They should be new and original or a radical adaptation of an old idea. As long as it is new and original, any field will be considered.

5. ARELOTT MFG. CO., INC.
4916 Shaw Avenue
St. Louis, Missouri 63110

Photographic, automotive accessories, marine accessories.

6. THE SHELBY METAL PRODUCTS COMPANY
110 Broadway, P. O. Box 525
Shelby, Ohio 44875
W. H. Kinnaird, *President*

To consider new products we require that the item be patented or that the inventor obtain and execute a disclosure form before submitting his idea or invention. We further require *one* element of compatibility from the following:

1. Principal components produced by metal stamping.
2. Principal market to be either hardware-electrics department of chain stores and or the builders hardware or housewares department of hardware and general merchandise wholesalers.
3. Principal market among industrial purchases either for incorporation in their product, cost reduction or sefety purposes.

Although our manufacturing processes are currently limited to metal stamping, assembly and electroplating we are not limited to these processes if a product has one of the other possible elements of compatibility. We are especially interested in items that will contribute to personal convenience, personal security or cost reduction in manufacturing.

7. THE McLAUGHLIN COMPANY
212 Jaikins Building
Birmingham, Michigan 48011
Robert B. Ryan, *President*

Manufacture of weld, pilot, and clinch nuts, special fasteners and bolt and nut assemblies.

We would consider purchasing inventions of fasteners for any industry.

The inventor should first write for a disclosure form before submitting his idea or invention, etc.

8. JEFFREY-ALLAN INDUSTRIES, INC.
2100 Greenleaf Street
Evanston, Illinois 60204
S. J. Kulwin, *President* and/or
Sy Bromberg, *General Manager*

Manufacturers of interior and exterior automotive accessories. Functional types mostly. Manufacturers of automotive safety belts and allied safety items. Distribution to normal aftermarket wholesale automotive distributors as well as to retail auto chain stores. International distribution or all our inventions. Primarily interests in those dealing with safety and for functional dressup accessories. A disclosure form is not necessary; we use our own.

Needed inventions: Improved side view mirrors-outside type. Improved defogging equipment for rear window. Improved top carriers for traveling. Improved means of defogging all windows inside. Improved winter items in following catagories; ice and snow scrapers, engine warmers for cold nights, device to eliminate getting stuck on ice and/or in ice ruts, improved devices on products for frozen locks, etc.

9. O. F. MOSSBERG & SONS, INC.
7 Grasso Avenue
North Haven, Connecticut 06473
Carl H. Benson, *Dir. R & D*

Firearms and sighting equipment for same. These are sold through various sporting goods distributors and mail order houses, etc.

We are interested in any invention pertaining to sporting arms and their accessory items. We are also open to any items to diversify our program.

We expect the inventor to fill out our disclosure form.

Needed inventions: Ideas pertaining to sports and recreation.

10. GENERAL SLICING MACHINE CO., INC.
Walden, New York 12586
H. S. Friedman, *Vice President-Sales*

General Slicing Machine Company is a well respected 35 year old company—national distribution in housewares industry. Also restaurant equipment, hardware, stationery, premium industries, sales promotional and executive gifts.

We sell home kitchen and commercial slicing machines (hand and electric), meat choppers, salad makers, vacuum-base products line including vises, vacuum-base pencil sharpeners, tape dispensers, vacuum-base ice crushers and knife sharpeners, etc.

To housewares departments of department stores, jobbers, distributors, premium outlets, executive gift companies, restaurant equipment dealers and stamp plans.

We would like to purchase inventions related to kitchen equipment —both home and commercial and all related pdts., particularly those suitable for mass premium markets.

Needed inventions: Items for food processing either/or home and commercial.

II.   UNITED SURGICAL CORPORATION
154 Midland Avenue
Port Chester, New York
Jack R. Blackwood, *President*

United Surgical Corporation, manufacturers and distributors of surgical products, instruments and specialties for forty-six years, enjoys world-wide prominence in the fields of surgery, medicine, and all branches of the medical profession.

United Surgical has the distinction of being the world's foremost manufacturer of "Ostomy" appliances and accessories worn and used by patients following radical intestinal and bladder surgery.

Needed inventions: Any product that improves the overall picture of public health in the form of a new product or the improvement of an old one.

12.   WHITMAN PUBLISHING COMPANY
1220 Mound Avenue
Racine, Wisconsin 53404
Miss Mary L. Hilt

We manufacture toys, playing cards, ribbons, gift wrapping, puzzles, and juvenile books.

Before we will examine any new item, "New Idea" submission forms must be filled out. We will send these on request. We do not examine any items unless this form is included.

13.   RAY GREEVE & CO., INC.
508 S. Byrne Road
Toledo, Ohio 43609
Mr. Ray Greeve, *President*

We manufacture boats and fiberglass parts, distribute through dealers and distributers, and are interested in general boat inventions. *Needed inventions:* Improved sailboat fittings.

14. DRAVO CORPORATION
Pittsburgh, Pennsylvania 15225
Dr. J. A. Anthes, *Manager of Research*

Products: Rivertow boats, harbor tugboats, barges, bulk materials handling equipment; ore processing plants for: Beneficiation (Concentration), Agglomeration, Smelting or other reduction; heavy concrete construction such as dams, locks, bridge substructures; power plants; chemical process plants.

Inventors having unpatented ideas or inventions must sign form before submitting. Ideas relating to above activities will then be considered.

15. FAIRCHILD CAMERA AND INSTRUMENT CORP.
300 Robbins Lane
Syosset, New York 11791
Mr. L. S. Smithers, *Patent Administrator*

The Fairchild Camera and Instrument Corporation manufactures primarily for the electronics industry. We are interested in reviewing any inventions pertaining to aerial camera systems; space navigational systems; photoelectric engravers; photocomposing machines; electrical resistors, transducers (pressure) etc.; electronic tubes, oscilloscopes, color motion picture cameras, television applications for military use, semiconductor integrated circuits.

A written disclosure or a patent should be submitted. Submitters are required to sign our standard inventions release form.

16. STEEL IMPROVEMENT AND FORGE COMPANY
970 East 64th Street
Cleveland, Ohio 44103
Mr. Ellsworth Smith, *Manager of Marketing*

Present products are a complete line of boiler manhole and handhole cover hardware for use on all types of boilers. Complete line of metallic ring gaskets for application in ring flanges in the piping industry. A line of rim clamps as used in the trucking industry. Our products are through distributors for replacement usually in the form of industrial or specialized types of distribution. We are interested in ideas related to the boiler and pressure tank field; also special truck and trailer chassis and applicable ideas in the valve and piping industry. Our policy does not require that an inventor first write for a disclosure form for submitting of ideas, but we do treat anything in that area in complete confidence.

A new economical low-cost air hydraulic landing gear with mechanical locking features. Landing gears to be used on trailers for transportation industry.

17. GENERAL BATHROOM PRODUCTS CORP.
2201 Touhy
Elk Grove Village, Illinois 60007
Jerry Epstein

Manufacturers of bathroom medicine cabinets, lighting fixtures and bathroom accessories.

Sell through building material distributors, electrical, tile, etc.

Would consider any invention that could be sold through these same channels of distribution.

18. TESCOM CORPORATION
2633-4 St. S. E.
Minneapolis, Minnesota 55419
D. R. Wedan, *Chief Engineer*

Tescom Corporation has two divisions. Smith Welding Equipment and Fluid System Division. The Welding Equipment Division manufactures a complete line of gas welding, heating and oxygen cutting equipment. Regulators for control of oxygen and fuel gases are also manufactured. The Fluid System Division specializes in high pressure (15,000 p.s.i.) regulators and valving equipment. Sale of welding equipment is through a number of independently owned welding equipment distributors. Sales of F. S. equipment is either direct or through manufacturing representatives.

19. Illinois Shade Division of the Slick Industrial Company
17th and Union Avenue
Chicago Heights, Illinois 60411
A. Arnold Parcels, *Technical Sales Eng.*

Our company manufactures window shades and plastic coated industrial fabrics. Using a base of woven or non-woven cloth, we apply diverse coatings for various specific end uses. Coated cloth is produced for shades, drapes, awnings, projection screens, art canvas, decking, wallcoverings, etc. The company can develop or "tailor" a coating for defined requirements. A disclosure form is required before any ideas are submitted.

Possible needed inventions may be in the area of new or improved coating machinery controls, methods or equipment. Any new or current product that requires coated cloth could be assisted by us. Consider a plastic coated cloth as a replacement for leather as a weather barrier or wear resistance.

20. DePUY MANUFACTURING COMPANY
P. O. Box 988
Warsaw, Indiana 46580
D. G. Scearce, *Chief Engineer*

The company manufactures and distributes through direct selling salesmen items of use to the hospitals, clinics, nursing homes, and in particular, to specialty fields of orthopaedic surgery, and general surgery. We make splints, braces, fracture frames, orthopaedic implants and instruments. We also buy and resell a similar line to compliment the items we make. Any idea associated with the general medical field and in particular the orthopaedic line, may be submitted, sketch or photograph, for our consideration. A general description of each item should accompany the disclosure. Do not submit sample unless requested.

21. VACO PRODUCTS COMPANY
510 N. Dearborn
Chicago, Illinois 60610

Manufacture hand tools, electrical connectors and miscellaneous fastener devices.
Distribution through distributors to professional mechanic.

22. SCHOELLHORN-ALBRECHT MACHINE CO.
721 North 2nd Street
St. Louis, Missouri 63102
J. J. Kiefer, *Manager*

We manufacture marine deck equipment for boats and barges.
Needed inventions: Inexpensive method of lashing barges to boat and to each other.

23. ECLIPSE FUEL ENGINEERING CO.
1100 Buchanan Street
Rockford, Illinois 61103
E. J. Skerkoske, *Regional Manager*

Manufacturer of industrial combustion equipment including gas and oil burners, valves, mixers, blowers and systems for industrial combustion applications. Distribution (sales) through representatives and factory branch offices.

24. ROTARY SEAL COMPANY
DIVISION/MUSKEGON PISTON RING COMPANY
7440 West Lawrence Avenue
Chicago, Illinois 60656
Edward F. Barrett, *Manager, Industrial Sales*

We are manufacturers of axial face-type seals for rotating or oscillating shafts. We sell both to the original equipment manufacturer through our direct field sales engineers and to the replacement market through our network of distributors. We currently ship seals to all parts of the free world. We would be primarily interested in shaft sealing devices, but would consider any product compatible with our manufacturing processes. No disclosure formalities are required.

25. STAS INSTRUCTIONAL MATERIALS, INC.
2100 FIFTH STREET
BERKELEY, CALIFORNIA 94710
Mr. Peter F. Young, *Consultant*

We design and manufacture scientific equipment and school teaching aids directed at the elementary and high school levels. We distribute our own products direct and also operate on a contractural consulting basis to several major publishing companies for the design and manufacture of manipulative materials required in their teaching programs.

We would be interested in inventions involving new and unique methods of teaching science and mathematics by the use of teacher demonstration apparatus and individual student manipulative materials.

We have no standard "red tape" procedures so an inventor may feel free to contact us in any form he wishes. Due to the normal legal problems involved in such contacts we do, however, require that he do not disclose his invention upon the first contact. This disclosure should be made through normally accepted procedures, which we will outline to each individual inventor.

26. KNAPE & VOGT MANUFACTURING CO.
2700 Oak Industrial Drive
Grand Rapids, Michigan 49505
James B. Vogt, *Vice President, R & D*

We manufacture builders hardware, including shelf standards and brackets, drawer suspensions, closet and kitchen accessories, sliding and folding door hardware and pegboard hooks.

27. SWEET MANUFACTURING COMPANY
Gilbert Street
West Mansfield, Massachusetts 02083
William R. Armstrong, Jr., *General Mgr.*

We are manufacturers of jewelry chain and jewelry findings which we sell to the manufacturer and wholesaler for distribution. We would be interested in inventions concern the fastening of chain such as on neckchains and bracelets. It is not necessary for the inventor to write for a disclosure form prior to submission of his idea.

28. E. R. WAGNER MANUFACTURING CO.
4611 N. 32nd Street
Milwaukee, Wisconsin 53209
Paul Speight, *Product Manager*

E. R. Wagner currently manufactures stamped metal components and assemblies, and heating elements for diversified industries including automotive, appliance, juvenile toy, lawn and garden equipment, etc. We review all inventions on a non-confidential basis only. All inventions must be covered by patent or disclosure before we will consider them.

29. MARDI GRAS ENTERPRISES
Lower Simms Street, P. O. Box 146
Simmesport, Louisiana 71369
Lucius Lacour, *Owner*

Small production and sales nationwide. Inventions we would be interested in are ideas that can be printed and made of paper and light cardboard, unique publications and booklets in the novelty field or gift. We do manufacture wood products, however, we are not interested in anything in this line unless it is simple and appealing and easy to make package.

30. THE UNION FORK AND HOE COMPANY
500 Dublin Avenue
Columbus, Ohio 43215
Mr. Walter Hart, *Chief Engineer*

We are engaged in the manufacture of farm, garden and industrial tools. We sell through distributors. Any new tool that could be used in the home garden, on the farm, or in industry (such as shovels, scoops. etc.) would be acceptable.

It will not be necessary to write for a disclosure form. Send all information and/or drawings to the attention of Mr. Hart, Chief engineer.

31. THE STANLEY WORKS
195 Lake Street
New Britain, Connecticut 06050
Walter R. Bush, *Director,*
Research and Product Engineering

The Stanley Works produces hand tools (essentially wood-working), electric hand tools, air tools, hardware, aluminum windows, drapery hardware, garage doors and operators, automatic doors and door controls, steel strapping, cold rolled steel, preservative paints, and several O. E. M. products. Our distribution varies with each market we approach. We are interested in inventions that could fit into any of the above product lines and, to some extent, inventions that might be companion product lines to the above. The inventor should first write for a disclosure form before submitting his idea even if a patent has already been applied for. In the event the product is patented, a disclosure form in advance is not necessary. Inventions that would be profitable are better ways to close door openings, better ways to clean shoes before entering buildings and more effective ways of working wood with power tools.

32. GLEASON CORPORATION
P. O. Box 343
Milwaukee, Wisconsin 53201
E. A. Smith

Basic manufacturers of:

(1) Heavy duty vehicle road safety equipment—Anthes Div. Heavy duty vehicle exhaust associated equipment—Anthes Div. Flares, fuses. Advance hazard warning (Highway)—Anthes Div.

(2) Wheels—plastic, solid rubber, semi-pneumatic—Gleason

Wheel Piv. Pneumatic.

(3) Casters.

(4) Hand Trucks—Portable manual material hauling equipment—Mil. Truck Div. Shopping carts, grocery market portable equipment—Mil. Truck Div.

Needed inventions: A simple, easily installed, dependable device that will alert the driver of a semi-trailer that he is losing air pressure in any of the truck tires, particularly trailer tires. This should be a fail-proof device without wires, brushes, etc. There is a tremendous market for such a device.

33. TOY INNOVATIONS, LTD.
350 Fifth Avenue, Suite 8011
New York, New York
Mr. George A. Spitzer

We furnish the wholesale and National Chain Grade with toys and games of various descriptions. Plastic, metal, wood, packaged on blistered cards and boxes.

34. LISLE CORPORATION
813 East Main Street
Clarinda, Iowa 51632
Mr. J. R. Arthur, *Vice President*

The Lisle Corporation manufactures mechanics speciality tools, such as Ridge Reamers, Cylinder Hones, Brake Hones, Ring Compressors, etc. We sell through automotive warehouses and automotive jobbers to garages and service stations. We are very interested in securing additional tools to add to our line. These should be fairly simple, low to medium priced. We are not interested in wrenches, hammers, pliers, etc., but in fairly universal speciality tools.

35. GEM ELECTRIC MFG. CO., INC.
390 Vanderbilt Motor Parkway
Hauppauge, Long Island, New York 11787
L. Pollan, *Sales Manager*

We have been in business approximately 39 years and manufacture over 1000 electrical items. Our wiring device line consists of household plugs, sockets, fuses; also some devices for industrial purposes, as well as cartridge fuses, and in addition, we produce a complete line of Christmas decorative lighting products. Always interested in new ideas; if they are feasible and have a potential, it would warrant the investment.

36. THE CARTER-WATERS CORPORATION
2440 Pennway
Kansas City, Missouri 64108
Albert R. Waters, *Chairman*

Our business consists of supplying materials to the construction industry, some of which we manufacture and some of which we distribute for other manufacturers. Many of our products have to do with concrete construction such as Hunt Process concrete curing compound, air-entraining agent for concrete, form oil for concrete forms, epoxy resins for various purposes such as floor topping, repairing spalled or cracked concrete, anchoring heavy machinery to concrete floors or piers. We manufacture "Haydite" which is a lightweight aggregate for concrete, also "Saturock" which is a cold mix asphaltic concrete that can be stored in a pile for future use or can be applied in almost any kind of weather.

As distributors, we handle face brick, glazed tile, reinforcing bars and many other products for construction.

37. JAK-PAK, INC.
P. O. Box 374
Milwaukee, Wisconsin 53201
S. J. Bernstein, *General Manager*

Manufacturer of toys for boys and girls—infant to 12. Toys are usually blister-packed for pegboard racks. Nationwide and some international distribution via a network of representatives. Toys should be simple in design, easily assembled or molded requiring a minimum of hand assembly.

Disclosure form preferred, but not necessary.

Needed inventions: Simple, relatively self explanatory, easy to produce, non-bulky toys.

38. ASTRO-DOME, INC.
1801 Browniee Ave., N. E.
Canton, Ohio 44705

Astro-Dome, Inc., manufactures observatory domes and planetarium equipment for the earth and space science field. We would be interested in reviewing related products in this field that can be sold through our agents primarily to the educational market.

39. ATLAS ASBESTOS COMPANY
North Wales, Pennsylvania 19454
William H. Johnston, *Vice President,*
General Manager

We are essentially weavers of industrial textile fabrics—primarily asbestos and/or fiberglass. One of these items has been sold to leading hardware distributors throughout the country with whom we have had good entree. We would be interested in ideas that would use our weaving facilities for consumertype products to be sold through hardware distributors, but we are not interested in gadgets and gimmicks.

40. COLOR CRAFT CORPORATION
P. O. Box 1651
Indianapolis, Indiana 46206
A. R. Van Wyngarden, *President*

Anodized aluminum housewares, *Shat-r-pruf* Glassware, some plastic molded items. We are basically manufacturers of aluminumware utilizing both stamping and drawing and spinning. We cover the entire United States and Canada with our sales force of both full time men and factory representatives, the wholesale and retail trade in housewares, the dairy industry with premium items and most of the major premium users. Items in related fields would be considered. The inventor can contact us direct anything submitted will be kept confidential. We would consider anything in the housewares field—especially those items in the kitchen gadget field.

41. ITT WIRE & ABLE DIVISION
95 Grand Avenue
Pawtucket, Rhode Island 02862
B. Wrubel, *Mgr. C. S. & W. D. Prod. Engr.*

We manufacture over 1000 types of wire and cable. Full line of all types fuses. Full line of trouble lights, wiring devices, cord sets, nite lights, switches, receptacles, plugs, etc. Distributed to manufacturers and super-jobbers and wholesalers through nationwide network of manufacturer's representatives. Some sales directly through to superchain stores. Facilities in Pawtucket, Clinton, Massachusetts; Woonsocket, Rhode Island, Burbank, California; Pt. Clare, Quebec. Inventions related to production line. No protection for unpatented ideas, although we would do our best.

Needed inventions: Almost anything involving cost or any idea on new wiring devices and related items that would attract impulse sales in our rack-jobbing program.

42. NATIONAL CYLINDER GAS (DIVISION)
CHEMETRON CORPORATION
840 N. Michigan Avenue
Chicago, Illinois 60611
Nicholas M. Esser, *General Patent Counsel*

We manufacture industrial gases in liquid or gaseous form. On-site industrial gas plants for large-volume users. Medical gases and mixtures. Welding and cutting supplies and equipment. Hospital piping systems and secondary equipment. Inhalation therapy and anesthesia apparatus and supplies. Railroad rail welding, cutting and maintenance equipment. Nitrogen food freezing and transport refrigeration systems. Air separation facilities are located in principal use areas.

43. CARDOX (DIVISION)
CHEMETRON CORPORATION
840 N. Michigan Avenue
Chicago, Illinois 60611
Nicholas M. Esser, *General Patent Counsel*

We manufacture gaseous and liquified carbon dioxide and dry ice. Carbon dioxide bulk liquid storage systems. Truck and truck trailer refrigeration systems. Chlorine dioxide hydrates. Foam system for airport fire trucks. Low pressure, high pressure and portable carbon dioxide fire extinguishing systems and equipment. Carbon dioxide producing units in eight strategic locations in the United States and manufacturing plant for fire extinguishing equipment at Monee, Illinois.

44. ALLOY RODS COMPANY (DIVISION)
CHEMETRON CORPORATION
840 N. Michigan Avenue
Chicago, Illinois 60611
Nicholas M. Esser, *General Patent Counsel*

We manufacture are welding electrodes in stick and continuous wire form for stainless and alloy welding. Flux-cored electrodes for welding low alloy and mild steel. Small-diameter copper-coated continuous wires and alloy-cored hardsurfacing electrodes for semi-automatic and automatic welding applications. Special electrodes for welding dissimilar metals.

45. THE McGEAN CHEMICAL CO. (DIVISION)
CHEMETRON CORPORATION
840 N. Michigan Avenue
Chicago, Illinois 60611
Nicholas M. Esser, *General Patent Counsel*

We manufacture a complete line of nickel anodes. Also brass, bronze, cadmium, copper, lead, tin and zinc anodes. Bright nickel plating salts and compounds. Nickel addition agents. Antimony, cadmium, calcium, chromic, cobalt, copper, iron, lead, manganese, sodium, tin and zinc salts, compounds and chemicals. Plants and laboratores at Cleveland and Detroit.

46. PENNSYLVANIA FORGE CO. (DIVISION)
CHEMETRON CORPORATION
840 N. Michigan Avenue
Chicago, Illinois 60611
Nicholas M. Esser, *General Patent Counsel*

We manufacture custom open-die forgings in unlimited size and shape. Complete heat-treating and machining facilities. Hollow-boring equipment for extreme lengths. Specialists for non-standard settings such as ells, tees, laterals and wyes, for high pressure and high temperature installations. Forged pipe flanges, standard or special. Custom forgings and flanges supplied in ferrous and non-ferrous materials to A. S. T. M., nuclear military and many other specifications.

47. ORGANIC CHEMICALS CO. (DIVISION)
CHEMETRON CORPORATION
840 N. Michigan Avenue
Chicago, Illinois 60611
Nicholas M. Esser, *General Patent Counsel*

We manufacture phosgene and phosgene derivatives, chemical and pharmaceutical intermediates, plastic additives. Plants at Newport, Tennessee; Elkton, Maryland; LaPorte, Texas.

48. TUBE TURNS CO. (DIVISION)
CHEMETRON CORPORATION
840 N. Michigan Avenue
Chicago, Illinois 60611
Nicholas M. Esser, *General Patent Counsel*

We manufacture welding fittings and flanges in a wide range of sizes, types and materials. Expander flanges, Sight glasses, Insulated joints, Pipe guides, Manual T-holt hinged closures, Automated hinged closures, Expansion compensators, Bellows expansion joints, Custom closed-die forgings. Plants at Louisville and Houston.

49. GIRDLER CATALYSTS CO. (DIVISION)
CHEMETRON CORPORATION
840 N. Michigan Avenue
Chicago, Illinois 60611
Nicholas M. Esser, *General Patent Counsel*

We manufacture nickel, chromium, palladium, platinum, cobalt, molybdenum, copper, iron and zinc catalysts. For hydrogen and synthetic gas manufacture, ammonia dissociation, sulfur removal hydrogenation, dehydrogenation, ammonia synthesis and purification of gas streams.

50. HOLLAND-SUCO COLOR CO. (DIVISION)
CHEMETRON CORPORATION
840 N. Michigan Avenue
Chicago, Illinois 60611
Nicholas M. Esser, *General Patent Counsel*

We manufacture a complete line of organic and inorganic flushed, dispersed and dry color pigments. Color intermediates. Vehicles, varnishes and driers. Pulps and press cakes. Extender pigments. Phonograph record compounds. Plastic traffic markings. Plants at Holland, Michigan; Huntington, West Virginia; Stockertown, Pennsylvania; St. Louis, Missouri.

51. VOTATOR COMPANY (DIVISION)
CHEMETRON CORPORATION
840 N. Michigan Avenue
Chicago, Illinois 60611
Nicholas M. Esser, *General Patent Counsel*

We manufacture continuous scraped-surface heat transfer equipment. High-speed thin film evaporators. Continuous and semi-continuous deodorizers. Rotary piston and gravity fillers. Hydrostatic sterilizers. Vacuum dryers. High frequency dielectric heating equipment.   Positive displacement pumps.   Controlled recycling mixers.

52. NORTHWEST CHEMICAL COMPANY (DIVISION)
CHEMETRON CORPORATION
840 N. Michigan Avenue
Chicago, Illinois 60611
Nicholas M. Esser, *General Patent Counsel*

We manufacture a complete line of industrial metal-cleaning compounds including, electro, soak, spray and solvent emulsion.   Etchants,
paint strippers, derusters, drawing and quenching compounds.   Iron
and zinc phosphating compounds and systems.

53. THE ALL BRIGHT-NELL CO. (DIVISION)
CHEMETRON CORPORATION
840 N. Michigan Avenue
Chicago, Illinois 60611
Nicholas M. Esser, *General Patent Counsel*

We manufacture meat processing, packing and allied industry equipment.   Slaughtering, cutting, curing, continuous and batch rendering,
slicing, packaging equipment, and all kinds and types of conveying and
handling equipment.   Complete cattle dressing systems.   Representatives in every important meat processing location in the world.

54. CLARK EQUIPMENT COMPANY
P. O. Box 31
Buchanan, Michigan 49107
K. C. Witt, *Chief Patent Counsel*

Our company manufactures automotive axles, transmissions and tor-
que converters for the automotive industry.   We further manufacture
off-highway equipment in the material handling field, construction
machinery, and transportation industry.   Further we manufacture a
line of refrigerated food display equipment.   Generally speaking we
would be interested in inventions that would advance the state of the
art in any of these areas.   The inventor should write for disclosure forms
before submitting his ideas unless the particular item has been used commercially or has patents issued for same.

Needed inventions: gyroscopic stabilizing devices for fork lift trucks,
novel power trains for rough terrain type vehicles, new or novel automotive power transmission devices, hydraulic accessories, electrical
accessories, flotation equipment.   refrigerating equipment,

55. BROWN & SHARPE MFG. CO.
Precision Park
North Kingstown, Rhode Island 02852
Peter E. Carbone, *Mgr., Design Engr.*
Joseph E. Kochhan, *Product Mgr. Industrial Products*

B. & S. produces machine tools, precision measuring instruments, cutters, hydraulic devices and machine tool accessories. We distribute our goods through authorized dealers and hardware stores.

We would consider purchasing inventions relating to precision linear measurements. An inventor does not need to request a disclosure form prior to submitting his idea or invention.

Needed inventions: Inexpensive, quick reading, hand-held measuring tool that may replace the micrometer caliper; materials having stability, unaffected by temperature.

56. AIRFAN ENGINEERING CO.
7401 Telegraph Road
Los Angeles, California 90022
David Reznick, *President*

We manufacture evaporative condensers, fan coil units, gas-fired multizone units, rotary evaporative coolers and blowers.

We distribute through local market by our salesmen, other markets through agents and are interested in purchasing inventions on heat transfer and air moving apparatus.

57. BUSH HOG, INC.
P. O. Box 1039
Selma, Alabama 36701
C. Price, *Proj. Engineer*

We manufacture the following: rotary cutters, disk harrows, planting equipment, rotary tillers, scraper blades, garden tractors, lawn and garden equipment,

We are interested in any inventions related to the above type implements.

58. COUNCIL MANUFACTURING CORPORATION
420 North Second Street, P. O. Box 243
Fort Smith, Arkansas 72901
Mr. Dansby A. Council

We manufacture icemakers, soft drink dispensers, bagged ice vendors, automatic ice dispensers, miscellaneous coin-operated devices for charcoal and similar products. We would be interested in inventions allied to our present equipment so that they might be sold through distributors and dealers.

59. ADVANCE CAR MOVER CO., INC.
112 N. Outagamie Street
Appleton, Wisconsin 54911
James F. Miller, *Manager*

Small firm manufacturing manual railway car movers. Hand spring coilers, hand melting ladles, hand trucks. Products are distributed by industrial and other distributors who sell to the end user—distributors throughout the USA and abroad. Interested in small bench mounted and/or hand tools, materials handling tools and devices. No disclosure form necessary.

60. AMERICAN HOSPITAL SUPPLY
2020 Ridge
Evanston, Illinois 60201
Tom J. Cowley, *Product Planning Mgr.*

Major distributor and manufacturer of products consumed by hospitals.

61. TEXIZE CHEMICALS, INC.
P. O. Box 368
Greenville, South Carolina 29602
Mr. C. R. Blumenstein, *Technical Director*

Texize is a manufacturer of textile chemicals founded in 1946 and today enjoying a business sales volume of approximately 28 million dollars. We look forward to a volume of 36 million dollars in 1967, which gives you some indication of company growth. Broadly speaking, the company is divided into three major sales areas: The consumer or household line which is concerned with soaps, detergents, cleaners, floor polishes, waxes, disinfectants, bleaches, starches, spot removers, etc. The industrial line is concerned primarily with industrial maintenance chemicals, paint strippers, floor polish sealants, anti-corrosion agents, metal processing chemical, germicidal and disinfectant compounds, etc. They sell broadly to the janitorial trade, the institutional maintenance trade, and the metals mechanical fabrications trade. The third divisional area is the textile chemical field which incidently is the basis of the company, having been founded as a sizing manufacturer and then branching out into textile finishing chemicals as well. We are interested in any new ideas applying to the above fields.

There is a need for an invention that would be a good prespotter for durable press polyester-cotton fabrics. There is always room for new germicidal compounds, germicidal cleaners and disinfectants.

Other needed inventions: A suitable dry floor wax stripper that does not entail the messy business of water and suds. A good rug shampoo or carpet cleaner that will allow a uniform non-spotted clean job on the floor without the messy matter of present day cleaners. A soil repellant finish that the housewife can put on to the fabric between washings to get "more mileage" out of garments between washings and launderings. A good truly wash and wax type of compound that will allow one to wash an automobile and put on a durable wax surface equivalent to a hard rub wax compound.

62. AMP INCORPORATED
Eisenhower Boulevard
Harrisburg, Pennsylvania 17011
Mr. Leon V. Whipple, *Advanced Planning New Products*

AMP manufacturers an extremely broad range of electrical terminals, splices, connectors and application tooling of all types for widely diversified markets around the world. Emphasis on new products has also created product lines of programming systems, "power packages" and other electrical-electronic devices.

63. XEROX CORPORATION
P. O. Box 1540
Rochester, New York 14603
Mr. Clarence Green, *Patent Department*
(Write to Mr. Green for a *Submission of Ideas* form.)

Xerox Corporation is a highly diversified, completely integrated enterprise for the research, engineering, development, production, and marketing of equipment and systems for graphic communications. The present thrust is on xerographic copying and duplicating machines for the office, engineering, and new products and services in related fields, such as facsimile recording.

In education, the company is concentrating upon the preparation and production of specialized instructional and library materials, in both conventional and microfilm formats.

A subsidiary, Electro-Optical Systems, specializes in exotic power, propulsion, and communication systems for the military/space market.

Inventions relating to any of these fields are constantly being sought, and independent inventors are welcome to submit their ideas for evaluation by patent experts.

64. NATIONAL DIE CASTING COMPANY
3635 W. Touhy
Chicago, Illinois
R. E. Johnson, *Vice President*

We manufacture and distribute almost all kinds of hardware, sporting goods, houseware, industrial products, using metal and plastic fabrications and assemblys.

65. BUNKHEAD MANUFACTURING COMPANY
P. O. Box 4
Houston, Texas 77001
S. J. Jamison

Sheet metal, light gage consumer products, distribution through wholesale grocers, hardware, sporting goods. Strongest distribution in south and southwest.

66. WIRT & KNOX MFG. CO.
23rd & York Streets
Philadelphia, Pennsylvania 19132
D. Weismuller, *Secretary*

We are manufacturers of fire fighting equipment, such as hand drawn hose carts, hose reels, racks, cabinets, etc. Our products are sold throughout the United States and overseas by distributors and jobbers of fire and safety equipment. We would be interested in inventions covered by our above type equipment.

Needed inventions: A method for keeping fire hydrants and piping systems thawed during winter.

67. AMETELS, INC.
East Moline, Illinois
F. R. Gruner, *Director of Engineering*

Ametels, Inc., manufacturers a wide line of industrial equipment, heavy machinery, gauges, instruments and controls. Some products are sold through agents, others by our own sales force. Inventions in the field of Chemical processing, centrifugals, filters, etc., laundry machinery (commercial), physical (metallurgical, plastics) testing machinery, instrumentation such as extensometers, etc., blowers and fans or other ideas in related equipment would be of interest.

Inventors must sign a disclosure form before submitting any information.

Needed inventions: An automatic clothes folding machine for home use. Commercial launderies, high speed feeder for larger ironers. Continuous laundry machine that is practical to do all washing and finishing. Improved air cleaning devices. Better equipment for liquid–solid and liquid–liquid separation in chemical processing.

In approaching the above problems studying the inventor might consider the Jet-Stream continuous laundry machine developed by Ling Temco, Texas.

68. EUREKA SPECIALTY PRINTING COMPANY (DIV.) LITTON INDUSTRIES
530 Electric Street
Scranton, Pennsylvania 18501
Mr. Lawrence E. Howard, *Engineering Director*

Company manufactures trading stamps, charity seals, coupons and other incentive merchandising aids. In addition, a major product area is in retail marketing systems, i.e. direct mail, catalogs, brochures, mail order forms and point-of-purchase displays. Distribution is nationwide.

Any and all inventions in the graphic arts, paper, adhesives, non-impact printing, laminating.

Needed inventions: A method of neutralizing static electricity, particularly in cold dry weather, which causes problems with feeding paper to printing presses. Remoistenable hot melt adhesives.

69. PULVERIZING MACHINERY DIVISION THE SLICK IND. CO.
Chatham Road
Summit, New Jersey 07901
E. L. Timm, *Technical Director*

Pulverizing Machinery Division manufactures industrial machinery used for the beneficiation of solids and fluids. Such equipment includes pulverizers, crushers, air separators, star feeders, screw feeders, rotary airlocks. We would be interested in improvements on these items plus new pieces of equipment in our field.

Our domestic sales are handled by manufacturer's representatives. Export sales through subsidiaries in England and Germany and licensees in Switzerland, Japan, Australia, and Argentine.

The inventor should not contact us until his attorney has made patent application and tells him it is alright to do so.

Needed inventions: An improved air washer filter to catch even the smallest dust particles (air pollution) to minimize maintenance problems in the power industry. Means of removing SO_2 gas from power plant stack effluent.

In approaching the above problems the inventor might consider that an air washer merely transfers the pollutants from the atmosphere to water. Water purification then becomes necessary.

70. POLYMER INDUSTRIES, INC. POLYMER SOUTHERN DIVISION
P. O. Box 2184
Greenville, South Carolina 29609
Richard E. Rettew, *Mgr. Development and Service Laboratory*

Basic products encompass emulsion polymers-vinyl acrylic and co polymers, reactions of natural polysaccharides involving starches and "gums" primarily. Above are distributed through marketing organization into textile field covering warp sizing, fabric finishing, printing and combinations. Industrial adhesives involving combining, paper and structures. Marketing in field is basically separated in textile and adhesive efforts. Markets are developed and serviced through market managers.

Inventions that could be of interest are those involving the product types above and market areas. Expansion into additional markets of allied nature is projected as industrial coatings, paper manufacturing, leather and related fields.

At this point company is primarily a manufacturing organization utilizing conventional equipment and concepts. By the same token new concepts are desired. These should be basic in nature rather than modification of existing techniques or compositions.

There is a need for an invention to apply desirable adjuncts to continuous structures from (1) anhydrous and (2) high solids, safe-economical solvent system (water considered solvent). Such application to be rendered durable with a minimum of energy expended. Polymers or prepolymers capable of reaction with fiberous substrates through graft type additions—sites to be those present in natural and most synthetic substrates. Methods of cleansing flexible organic structures (fabrics, films, leathers, paper like items, etc.) that require neither water nor organic solvents. Methods and techniques for point point bonding of fibrous structures by energy only, Methods and techniques for point–point bonding chemical reactant–energy.

The inventor might consider the above problem in term of separate application of co-reactants with subsequent making use of energy forms other than heat and radiation for accomplishing reaction. Utilization of co-reactants, one of which could act as solvent for the prereacted state and be consumed in final reaction.

71. RIVERSIDE CHEMICAL CO., INC.
River Road
N. Tonawanda, New York 14120
Dr. C. H. Rasch

Needed inventions: A construction material or extender that would cut road construction costs inhalf; methods to reduce air and water pollution; methods for the disposal of municipal garbage and debris; better methods for maintainence of highways and roads; better methods for marking highways and roads, better methods for picking up leaves in city streets; better methods for snow removal from highways and parking lots.

72. J. H. WILLIAMS & CO.
400 Vulcan Street
Buffalo, New York 14207
J. R. Haynes, *Plant and Product Engineer*

Manufacturer of hand tools and special forgings. Distribution: industrial supply houses. Inventions: Hand Tools.

Needed inventions: A non-slip universal grip wrench that can be used on any irregular sized nut.

73. UNIVERSAL VISE & TOOL COMPANY
8500 E. Michigan Avenue
Parma, Michigan 49269
Clark C. Chandler, *Vice President*

We presently manufacture work holding and positioning tools; air and hydraulic cylinders; standard drill fixtures and locks and a vertical spindled surface grinder. We distribute nationally through wholesalers and representatives. We would strongly consider the purchase of any type invention that would be complimentary to our existing lines.

Needed inventions: A universal jaw or device for a vise to secure odd-shaped articles.

74. SWINGLINE, INC.
3200 Skillman Avenue
Long Island City, New York 11101
Benjamin B. Hampton, *Vice President*

Swingline, Inc., manufactures mechanical and electrical staplers, mechanical and electrical pencil sharpeners, rubber bands, tackers, corrugated carton staple extractors and, in conjunction with its subsidiaries, industrial stapling equipment, riveting guns for the *Do-It-Yourselfer*, chemical adhesives and fasteners for the automotive trade, etc. We would consider any invention or modification of those items in our line. An inventor must first request a disclosure form before submitting his invention.

75. LILLISTON CORPORATION
P. O. Box 407
Albany, Georgia 31702
William G. Moore, *Vice President,
Engineering*

Manufacturer of quality farm equipment in a modern facility (completed October 1967), ten miles west of Albany, Georgia, on U. S. Highway No. 82. Presently producing three product lines, as follows: (1) Rotary cutters, eight models, tractor drawn and tractor powered; (2) peanut harvesting equipment. (3) cultivating equipment, four sizes of rolling cultivators.

Distribution through six branch warehouses to franchised farm equipment dealers in 19 states; through recognized farm equipment distributors in the remaining states; through Canadian subsidiary in Canada; through an export agent to all countries of the world.

Inventions that would be of interest to us would include all types of agricultural equipment that would help the U. S. farmer.

76. UNIVAC DIVISION
SPERRY RAND CORPORATION
P. O. Box 8100
Philadelphia, Pennsylvania 19101
Submission Section: Patents and Licensing

Interested in: electronic digital computers and related peripheral equipment, including circuits and/or mechanisms therefor and various uses‾of such equipent.

Needed invention; A system for sorting mail for distribution.

77. STERLING PLASTICS COMPANY
Sheffield Street
Mountainside, New Jersey 07092
P. E. Bauldry, *Vice President*

Manufacturer of injection molded plastic school and stationery supplies. Sell through wholesalers to all types of retailers. Interested in any ideas that can be used by students or offices and can be sold in volume through above outlets.

78. RUSSELL MANUFACTURING COMPANY
999 Liberty Road
Lexington, Kentucky 40501
Mr. Russell Hicks, *President*

We manufacture pocket knives, manicure files, seam rippers and carton cutters.

Needed invention: Improved desk-type calendar for home and office.

79. McGILL MANUFACTURING COL, INC.
ELECTRICAL DIVISION
N. Campbell Street
Valparaiso, Indiana 46383
J. S. Eason, *Vice President and
General Manager*

Manufacture electrical wiring devices such as, switches, portable extension lights, lamp changes, sockets, wire lamp guards.

Products are sold through authorized electrical wholesalers and direct to original equipment manufacturers.

Would be interested in inventions concerning electrical products that could be sold through the above listed channels.

80. UNITED STATES SAFETY SERVICE CO.
1535 Walnut
Kansas City, Missouri
N. G. Friesenborg, *Exec. Vice President*

Manufacture and sell industrial safety equipment; specializing in the fields of head and eye protection.
We sell direct to industry through our own sales department.

81. EMPIRE PRODUCTS, INC.
9201 Blue Ash Road
Cincinnati, Ohio 45242
Robert Senior, *President*

Manufacturers of plugs and receptacles and connectors. Also various power distribution devices. Sales direct to heavy industry and also distributor and manufacturing distributors.
We are interested in all types of patents relating to electrical power distribution devices.

82. WELLS LAMONT CORPORATION
6640 West Touhy Avenue
Chicago, Illinois
Carroll G. Wells, *Vice President*

All types of work and dress gloves, 9 factories, 2000 employees. Distribute through jobbers and chain stores and large department stores. Nationally from four distribution warehouses. North, East Coast, West Coast and South.
Needed inventions: New type sewing machines for fabrication fabric and leather gloves. A power semi-automatic glove turner for turning leather and combination leather and fabric gloves from inside out to right side out.

83. PARKER METAL GOODS CO.
85 Prescott Street
Worcester, Massachusetts 01605
Pern A. Consigcio, Jr., *Vice President*

We manufacture wire forms, metal stampings, schopping carts, household hardware, injection molding, TV antennas and hardware, bar stools.

84. N. A. WOODWORTH CO.
1300 E. Nine Mile
Detroit, Michigan 48220
George Hohwart, *Vice President,*
Production Department

We manufacture precision work holding devices such as chucks-arbors, etc. We have an internal sales organization. We are interested in workholding devices, jigs, gages.

Inventor should have his ideas secured before disclosure, i.e. patent applied for or notarized and witnessed.

85. STA-RITE GINNIE LOU, INC.
Shelbyville, Illinois 62565
G. Noel Bolinger, *President*

Manufacturer and distributor of hair care products, including bobby pins, hair pins, barrettes, hair nets, caps, clips, bonnets, curlers, bandeaus, hair bows, combs and brushes.

86. OLDERMAN BRASS MANUFACTURING CORP.
P. O. Box 917
Bridgeport, Connecticut 06601
Mike Olderman, *President*

We manufacture plumbing specialites, hardware specialties and marine hardware. Our trade: wholesale distributors and manufacturers. Inventions interested in: The above mentioned. forward inventions for our consideration and approval. Our facilities: Foundry for non-ferrous casting, automatic screw machines and secondary machine operation, plating equipment and assembling.

87. ROME PLOW COMPANY
P. O. Box 48
Cedartown, Georgia 30125
C. C. Mullen, *President*

We are manufacturers of heavy-duty harrows, deep tillage equipment and land clearing attachments for agricultural and Industrial tractors, motor graders and other prime movers. We would be interested in examining only those inventions after the patents have been applied for.

88. REMACO, INC.
200 Paris Avenue
Northvale, New Jersey
J. Wm. Muino, *Vice President*

Complete line automotive aftermarket. Distribution nationally and worldwide. All patents considered. Present interest is in tires and tubes.

Problem to be solved: what to do with scrap used autotruck tires (they cannot be burned)—to use them for a secondary product or develop a method of disposal.

89. GABRIEL INDUSTRIES, INC.
200 Fifth Avenue
New York, New York 10010
A. F. Nordstrom, *Product Development Mgr.*

Diversified manufacturer of toys and sporting goods.

90. THAYER, INC.
205 School Street
Gardner, Massachusetts 01440
Philip J. Carney, *Treasurer*

Manufacturers of juvenile products, such as cribs, cases, nursery groups, cribmobiles, play pens (both wooden and net), carriages, strollers, walka bouncers, high chairs, nursery chairs, car beds and car seats, and mattresses and pads. We also manufacture children's rockers, and doll high chairs and doll cradles in our toy line.

91. MOORE PIPE & SPRINKLER CO.
P. O. Box 3037
Jacksonville, Florida 32206
J. B. Evans, *President*

We manufacture automatic sprinkler systems. We have sales and engineering offices throughout the southeast. Interested in new inventions concerning fire protection and associated equipment.

92. ART-PHYL CREATIONS
508 Frelinghuysen Avenue
Newark, New Jersey 07114
Art Hochman

We manufacture display and merchandising aids.

93. THE METAL WARE CORPORATION
1710 Monroe Street
Two Rivers, Wisconsin 54241
E. T. Christoffel, *Product Engineer*

We manufacture electric housewares such as coffee makers, broilers, corn poppers and hot cups. Also electric toy ranges and toy corn poppers. Our sales are handled through sales representatives. We would consider any invention that would fit in with the above products.

We do not supply disclosure forms, but we will look over a patent and return it with reasons for not accepting it. Many of the patents we have examined have been too complicated and impractical for mass production.

Needed inventions: Most kitchens are too small to store a lot of electric household cooking appliances. A central storage area with space for toaster, broiler, corn popper, coffee maker would be desirable. Combinations of two and three appliances have been made but have not been successful, e.g. bread toaster, grill and broiler in one.

A family of appliances as mentioned above that would be easy to locate and store when not in use would be very useful.

94. S. M. FRANK & CO., INC.
18 East 54th Street
New York, New York 10022
Samuel M. Frank, Jr., *President*

S. M. Frank & Co., Inc., manufacturers of smoking pipes and certain other smoker's articles under the trademarks, Kaywoodie, Medico, Yello-Bole and several others. Primary distribution is through wholesale tobacco distributors and through direct sale to large retail chain drug stores.

The company is interested in any type of article relating to the smoking pipe and smoker's articles field, as well as any other types of articles that might possibly be sold through the above mentioned avenues of distribution. It is not necessary that the inventor write for disclosure form but all inventions must be submitted without any obligation on the part of the company.

Needed inventions: Anything emphasizing convenience for the pipe smoker.

Note to Businessmen

We would like to list your firm in future editions of this book. Write *Inventor's Handbook*, 4447 Lafaye Street, New Orleans, Louisiana, giving a brief description of your company and the types of inventions you would consider purchasing.

Index of Foregoing Companies

(Products they presently manufacture, method of distribution, and types of inventions they would consider purchasing. Numbers correspond to those in preceding list of companies.)

Companies Interested in Financing Inventions

The following companies are interested in financing promising inventions. We would suggest that an inventor contact these companies only after a successful prototype model has been built or an experiment run so that they would be in a better position to judge the worth of an invention.

The financial help and assistance could be in the form of development or setting up a corporation to develop and market the invention and the purchase of some of the corporation stock or in the form of a loan with the invention as collateral, etc. These firms often will have the experience of where and how to commercialize the invention.

Battelle Development Corporation
505 King Avenue
Columbus, Ohio 43201

American Research and Development Corp.
200 Berkeley Street
Boston, Massachusetts, 02116

Adelphia Capital Investment Corp.
Room 807
1518 Walnut Street
Philadelphia, Pennsylvania 19102

Arthur D. Little, Inc.
Acorn Park
Cambridge, Massachusetts 02140

The submission of inventions from outside sources is welcomed by Arthur D. Little, Inc. Because it recognizes that individuals and corporations cannot always properly commercialize their own inventions, ADL can under a variety of special circumstances provide technical and commercial assistance at no cost to invention owners.

The following outlines ADL's policies on the submission of inventions and the screening criteria that are used for their evaluation. It also suggests the type of invention projects in which ADL is interested and the basis on which co-operative arrangements may be established.

Technical requirements

ADL does not serve merely as an invention broker, but normally expects to contribute to the technical or commercial development of the invention, or both. Inventions submitted to ADL should represent substantial technical achievements; they should not be merely minor improvements, design modifications, or styling changes. Although the invention need not be well developed, the submitter should be able to give evidence of its technical soundness.

The submitted invention should stimulate the enthusiasm of the staff members responsible for the evaluation and continuing work at ADL. Since ADL's staff members deal largely with advanced technology, it is not probable, although it is possible, that they would be interested in such items as household devices, automotive accessories, toys, games, and wearing apparel.

Market considerations

The invention should be one that would ultimately be purchased by a reasonably large number of non-government buyers.

The attainable market for a potential new product should be at least one million dollars per year. A new manufacturing process should permit savings of at least several hundred thousand dollars per year.

Economic factors

If the invention is an improvement of existing products or processes, the inventor should submit sufficient data to indicate that it is economically competitive with the existing products or processes. If the invention cannot be compared to existing products or processes, the inventor should, in so far as possible, provide any information useful in assessing the economic aspects of the invention.

Legal considerations

The invention must be the subject of a United States patent or patent application, and should be sufficiently novel to offer reasonable protection to a licensee.

Licensability

The licensability of an invention depends upon the scope of its patent protection, the cost of implementing a licensing program, and the difficulty of enforcing the patent. ADL prefers inventions that can be licensed exclusively or to a small number of companies. It is not interested in situations where substantial patent infringement is believed to exist.

In general, ADL expects the submitter to give a complete summary of any prior efforts at commercialization or licensing, including the names of parties approached.

Disclosures to ADL

Unless ADL consummates a formal agreement with a submitter, the rights of the submitter will be only those which are afforded by the patent laws, and the rights of ADL will include any rights which the public may have. ADL will use due care to prevent disclosures from being communicated improperly to others, and will treat material in accordance with its established professional standards.

The disclosure must include a copy of either a United States patent or an application for a patent. It is not necessary for the submitter to disclose either the claims or the filing date, should he submit a patent application. Any submitted patent should be no more than five years old.

ADL can received submitted information only from the invention owner or from his authorized representative. It is not free to receive information from brokers or from individuals who may require a finder's fee.

Every submitted invention should be accompanied by a signed copy of ADL's standard Disclosure Form, which is available on request. While samples and models are not required initially, the inventor should amplify the information contained in the patent or patent application to whatever extent is necessary for a complete and accurate description of the invention to be made. The disclosure should include information on the availability of test data, samples, prototypes and the like.

Upon receipt of the disclosure, appropriate staff members or consultants regularly used by ADL will make a preliminary evaluation of the invention's technical merits and commercial potential. Usually ADL can advice the submitter within two to four weeks whether it is interested in further investigating the invention. If the results of the preliminary review are positive, arrangements will be made to meet with the inventor for the purpose of obtaining additional information.

If it appears that ADL can assist the inventor, it will propose an arrangement defining the commitment of both parties, whereby it agrees to do whatever it feels is necessary to establish a profitable licensing situation, in return for a portion of the licensing proceeds. Normally this arrangement will include a grant to ADL of the exclusive right, for a limited period of time, to develop and license the invention. While many agreements are based upon an equal division of licensing revenue, ADL considers the factors in each situation before suggesting a formula for income sharing.

NOTE TO INVESTORS:

We would like to list additional firms willing to finance inventions in future editions of this book. Write: *Inventor's Handbook*, 4447 Lafaye Street, New Orleans, La. 70122

Inexpensive Invention Review and Evaluation

The following is a list of scientists and companies willing to review inventions for inventors and to perform research work for hire. The review will be an opinion as to the technical soundness of the idea and to locate any obvious fallacies in principles or known scientific theory and to identify any outstanding obstacle to commercial success. Where possible, an opinion as to the actual need for the invention will be given and recommendations for either abandonment or suggestions of where the inventor should proceed. If the reviewer thinks the idea has promise, then, where possible, a price will be given to test out the idea either by a simple experiment or the construction of a crude working model.

As a token payment for this extremely valuable service, we require that each invention or idea for an invention to be reviewed be accompanied by a check for ten ($10) dollars made out to the reviewer.

Mr. Michael W. Pastore
Scope Associates
P. O. Box "C"
West Simsbury, Conn. 06092
Mechanical, heat transfer, heating and air conditioning, fuel control, stress dynamics, electromechanical.

Mec-Tron Corporation
239 Wisconsin Avenue
Racine, Wisconsin 53403
Electronics, Electromechanical, Mechanical

Parkway Products, Inc.
1230 West Seventh Street
Cincinnati, Ohio 45203
Plastics processing, model making, master tooling, cost analysis, mechanical design, metals engineering.

Scientific Control Laboratories, Inc.
3136 South Kolin
Chicago, Illinois 60623
Electroplating, electropolishing, bright dipping.

Philip E. Tobias Associates
2537 Mount Carmel Avenue
Glenside, Penna. 19038
Graphic arts, printing, photography, copying, chemistry, electronic and mechanical design.

Automation Engineers, Inc.
344 West State Street
Trenton, New Jersey 08618
Data processing

Mr. William Parker
32 Westminster Street
Worcester, Mass 01605
Mechanical, chemical, electrical

Applied Science Laboratories, Inc.
135 North Gill Street
State College, Penna. 16801
High purity lipid chemicals, gas chromatography, thin layer chromatography

Mr. Alf Hundere
Alcon Aviation, Inc.
P. O. Box 28299
San Antonio, Texas 78288
Internal combustion engines (mainly aircraft) lubricants and mechanics, fuels

Wilson Laboratories, Inc.
P. O. Box 9851
6985 Market Avenue
El Paso, Texas 79989
Inorganic chemistry, organic chemistry, colloid chemistry with special applications in agricultural chemistry.

Ted Wolf Associates, Inc.
387 Passaic Avenue
Fairfield, New Jersey 07007
Machine design, product design (mass produced consumer products), plastic parts and tools. Models and prototypes.

American Plastic & Chemical Corp.
Northboro Road
Industrial Park
Marlboro, Mass.
Resins, coatings and plastic additives.

We would like to list additional qualified reviewers in other fields, particularly reviewers able to carry out simple experiments to test the principles of inventions submitted. Write: *Inventor's Handbook*, 4447 Lafaye Street, New Orleans, La. 70122.

Inventors Willing to Collaborate in Developing Your Inventions

The following is a list of inventors willing to help you in developing and perfecting your invention. If this person proves valuable, suitable mutually satisfactory contracts should be arranged privately for a development program:

Mr. C. Gordon Haupt, F.A.I.C.
P. O. Box 4353
Roanoke, Va. 24015
Compounding and evaluation of chemical materials—plastics, metallics, water, electroplating, epoxy and pesticide.

Mr. Charles Medeiros
P. O. Box 221
Allston, Mass. 02134
Mechanical abilities

Mr. Milton Meyers
2918 Clinton Avenue
Fort Worth, Texas
Mechanical and some electronics

Mr. Lloyd D. House
RD #3
Montrose, Penna 18801
Tool die—model maker

Mr. Bob J. Bussoli
213 Montgomery Avenue
Cle Elum, Washington 98922
Toys, houshold items, gadgets

Mr. W. Preston Haupt, C.D.P.
114 Taylor Street
Chincoteague, Va. 23336
Chemical specialties, moire applications, aerosol safeties, automatic data processing, special slide rules, mechanical linkages and oils.

Mr. Clyde H. Farmer
200 C Street
Beckley, W. Va. 25801
Machine, machinist, hydraulic

Mr. E. C. Carter
P. O. Box 682
Rio Grande City, Texas 78582
Toys, hand and electrical

Mr. Wendell W. Pingel
Rural Route
Wabeno, Wisconsin 54566
Hunting, fishing equipment, automotive accessories

Mr. Justin Tieri
628 Parkside N.W.
Grand Rapids, Mich.
Model maker, mechanical and development ability

Mr. M. H. Friedler
14 Orange St.
Lewiston, Maine 04240
Cardboard and fibre products for the dry-cleaning trade and hat manufacturing—hat boxes

Mr. Jacob Cooper
24 Talbot Street
New Hyde Park, N.Y. 11040
Watchmaker

Mrs. Carla R. Rittenhouse
729 17th Ave.
East Moline, Ill. 61244
Artistic designing in mechanical, novel and gift items, also medical ideas.

Mrs. Rebecca Cohen
3015 Brighton 3rd St.
Brooklyn, N.Y.
General inventive interests

Dr. Samuel Klein
353 Hawthorne Ave.
Newark, N.J. 07112
Analysis and development of foods and medicines

Mr. Jack Huddleston, AIA
2210 Line Ave.
Shreveport, La.
Architectural, engineering, artistic

Mr. Daniel L. Bechtel
P. O. Box 11281
Fort Worth, Texas 76110
Firearms, gunsmith tools, ballistics

Mr. Charles L. Cooper
704 Grand Drive
Metairie, La. 70003
Equipment and machinery development—fabrication and installation.

Mrs. J. C. Toth
2045 Spring Road
Cleveland, Ohio 44109
General inventive interests

Mr. Jerry Marshall
738 W. 17th St.
San Bernardino, Calif. 92405
Model maker, drawings

Industrial Models
P. O. Box 39
Rome City, Indiana 46784
Prototype models—electrical, mechanical.

Mr. J. C. Downey
850 Maury Road
P. O. Box 117
Orando, Florida 32804
Plastics, adhesives

Mr. J. C. Muirhead
898 - 13th St. N.E.
Medicine Hat, Alberta, Canada
**Air flow (high and low speeds) force and pressure measurements,
   optics, photography (high speed)**

Mr. W. A. Sanzenbacher
P. B. 341
8022 Zurich, Switzerland
Mechanical

Mr. R. K. Anderson
R.D. #1
North Benton, Ohio 44449
Complete machine shop, electro mechanical design experience

Mr Nello J Orsini
233-W-11-Mile Road
Madison Heights, Michigan
**Toys, games, auto accessories, home products, various mechanical
   products.**

Dr. Joseph Dollinger, M.D.
8223 Bay Parkway
Brooklyn, N.Y. 11214
Medical

Mr. Thomas R. Yess
156 Shrewbury Street
Worcester, Mass 01604
Drafting, tool designing, machine shop, welding, brazing, plating, finishing.

Mr. Dal Creighton
1504 Crescent Place
Lakeland, Florida 33801
Mechanical, electrical, electronic, metal and plastic.

Mr. Jim Clary
P. O. Box 406
Benicia, Calif 94510
Casting in the non-ferrous metals

NOTE TO INVENTORS:

We would like to list your name as available to collaborate with other inventor's in perfecting their inventions, realizing that one day you may need someone's help in perfecting your own. Write: *Inventor's Handbook*, for 4447 Lafaye Street, New Orleans, La. 70122

Model Makers

The following is a list of model makers obtained from a number of sources.

ELECTRO MECHANICAL LABORATORIES, INC.
102 Westport Avenue
Norwalk, Connecticut 06851
Prototype instruments and short-run production

ENTERPRISE MACHINE & DEVELOPMENT CORP.
100 Fernwood Avenue
New Castle, Delaware 19720
Prototypes

GENERAL TOOL OF KALAMAZOO, INC.
615 W. Ranson at Lawrence Square
Kalamazoo, Michigan 49007
Prototypes

AUTO AIR PRODUCTS, INC.
S. Holmes Street
Lansing, Michigan 48912
Wood and plastic models

PARK WAY PRODUCTS, INC.
1232 W. 7th Street at McLean
Cincinnati, Ohio 45203
Engineering, scale models, prototypes

CENTRAL MODELS, INC.
112 W. Hubbard St.
Chicago, Illinois 60610
Models in any field

CRAFTOY MODELS
3605 Riverside Drive
Chicago, Illinois 60648
Electromechanical models, prototypes

INDUSTRIAL MODELS & PROTOTYPES, INC.
2025 W. Churchill St.
Chicago, Illinois 60647
Metals, wood, plastic models

INDUSTRIAL PATTERN MANUFACTURING CO.
3001 N. Oakley St.
Chicago, Illins 60647
Samples, prototypes—plastics, metals, wood

MODEL BUILDERS, INC.
6155 S. Oak Park
Chicago, Illinois 60636
Product models, precision prototype parts

BLOOMFIELD MACHINE CO., INC.
Bloomfield, Connecticut 06002
Prototypes

JOHNSON MANUFACTURING CO.
54 Brothwell Street
Bridgeport, Connecticut 06605
Prototypes

ABEX ENGINEERING LABORATORIES
486 Hammock Point Road
Clincon, Connecticut 06413

RADIAD SERVICE, INC.
8243 Elmwood Avenue
Skokie, Illinois 60076
Custom design and manufacturing

MOYER & CO.
2157 W. Division St.
Chicago, Illinois 60622
Experimental model work

OMNI MODEL & PLASTIC, INC.
548 N. Hyde Park Hillside
Chicago, Illinois
Mock-ups, prototypes

PRECISE DEVELOPMENT ENGINEERS
1767 W. Armitage St.
Chicago, Illinois 60622
Models for any purposes

STRICKER - BRUNHUBER CORP.
21 W. 24th Street
New York, New York 10010
Mechanical models

TOLEDO METAL SPINNING CO.
1823 Clinton Street
Toledo, Ohio 43607

PRODUCTION PREVIEWS, INC.
29 E. 21st Street
New York, New York 10010
Engineering technology for product development

MR. L. E. VAN HALST
Rome City, Indiana 46784
Industrial models, electrical, mechanical

MR. WILLIAM PARKER
32 Westminster
Worcester, Massachusetts 01605
Models, manufacturing, research

MILBURN
Burlington, Kentucky 41005
Small lot manufacturing in metals, plastics

The National Bureau of Standards, Washington, D. C., will for a fee do calibrations, and electrodeposition, and mechanical, optical, electrical and other measurements. Also a list of commercial laboratories available for hire is to be found in *Directory of Commercial and College Laboratories*, Superintendent of Documents, Washington, D. C. or *A List of Small Business Concerns Interested in Performing Research and Development*, U. S. Small Business Administration, Washington 25, D. C.

Services for Inventors

(Washington, D. C. and surrounding area)

Patent Draftsmen

Baron, John R. Associates	Warner Building
Borland, Eugene V.	1507 M Street, N. W.
Dueno, Hernani	10 Arthur Drive, Dixon Hall
Everett, Henry B.	1329 E Street, N. W.
Frederick, Andrew E.	1329 E Street, N. W.
Keithley, Howard W.	National Theatre Bldg.
Lestoevel, Pierre, J.	Munsey Building
Munger, Ormond S.	5500 Glenwood Road
Myers, Geo. A.	National Press Bldg.
Strauss, Harold	815-15th St., N. W.
Walsh & Associates	501-13th Street, N. W.

Patent Reproductions

Bocarsilski, Martha E.	1224 H Street, N. W.
U. S. & foreign patent drawings, designs, trade marks, court exhibits, offset printing	
Kirby Lithographic Co., Inc.	409-12th Street, N. W.
Foreign design and U. S. dimensional drawings	
Patent Reproduction Co.	26 N. SE
U. S. & foreign patent drawings	
Quinn Patent Drawing Service	1020 3rd, N. W.

Patent Searchers

Abstracts of Patent Service	Munsey Bldg.
Beckman, Kenneth W.	Munsey Bldg.
Bond & Altman	Munsey Bldg.
Capitol Patent Service	Munsey Bldg.
Dick, Talbert M	Munsey Bldg.
Hastings, Ann, Miss	711-14th St. N. W.
Hurson, Jos. V. Associates	1026-17th St. N. W.
Invention, Inc.	Munsey Bldg.
Mitrivar	1401 Wilson Blvd., Arlington
O'Brien, Jos	711-14th St. N. W.
Patent Engineering Service	7111-14th St. N. W.
Smith, Raymond L.	711-14th Street, N. W.
Stan Stanton's Search Service	927-15th St., N. W.
Wagner, Richard S.	Munsey Bldg.

Patent Services

Parker and Parker   3703 Huntington, N. W.
Foreign patents and trademarks

Patent Law Secretarial Service

Specializing in complete service for   605-14th St. N. W.
  patent attorneys, stenography, du-
  plication, legalization

Patent Translators

Charlotte Translating Bureau   8085 Beach Tree Rd., Bethezda, Md.
Geiger Associates   2025 I Street, N. W.
Institute of Modern Languages, Inc. 1666 Connecticut Ave., N. W.

Patent Developers

Creative Industrial Arts   630 S. Picket, Alexandria, Va.
Electrical, electronic and mechanical
  design and development
Dock & Dock   Munsey Bldg.
Commercial Patent Counsel
Evans, Jos. D.   1319 F Street, N. W.
Modern Machines of Va.   103 E. Sager Ave., Fairfax, Va.
O'Brien, Jos.   711-14th Ave.
Thorion, Inc.   8316-20th Ave., Hyatts

(New York)

Patent Development and Marketing

A Model Invention Development Co.   Woolworth Bldg.
A Patent Drafting Service   Woolworth Bldg.
Adams, Gilbert   15 Park Row
Artmen, Leonard H. & Associates
Capitol Industries, Inc.   44 E. 52nd
Design Patent Service   1328 Broadway
Dock & Dock   26 Broadway
Flexsleev, Inc.   104-5th Ave.
Foreign Operations Service, Inc.   333 E. 46th St.
Futuraire Development Corp.   20 Jerusalem Ave., Hicksville, N. Y.
  Inventions developed and model
  work
Heberlein Patent Corp.   350 Fifth Ave.
Interstate Mechanical Labs, Inc.   427 W. 51st St.
Israel Patents Corp.   170 Broadway
Lawyers and Merchants Translation   11 Broadway
  Bureau
Lee Raymond   130 W. 42nd Street
Leeds & Micallief   160 Fifth Avenue

Market Potential Corp.	969 Third Avenue
Mercury Drafting Service, Inc.	5 Beekman
Patent Drafting Corp.	11 E. 44th
Patent Exchange	26 Broadway
Patents Bought and sold	
Resources and Facilities Corp.	100 E. 42nd St.
Schwartz, M.A.	118 E. 28th St.
Spector, George	Woolworth Bldg.
Technical Lithographers, Inc.	118 E. 28th St.
Reproductions of patent and trade-mark drawings for U. S. and foreign filings	
Product Finder Co., Inc.	210 Broadway
Western Management Consultants	555 Madison Ave.

INDEX

www.ingramcontent.com/pod-product-compliance
Lightning Source LLC
Chambersburg PA
CBHW021030210326
41598CB00016B/967